Dictionarium
Bibliothecarii Practicum
ad usum internationalem in XXII linguis

The Librarian's
Practical Dictionary
in 22 languages

Wörterbuch
des Bibliothekars
in 22 Sprachen

Edited by / Herausgegeben von
Dr. Zoltán Pipics

6th revised and enlarged edition / 6. verbesserte und erweiterte Auflage

VD 1974
Verlag Dokumentation, Pullach / München

Collaborator / Mitarbeiter: Dr. József Bödey

Die erste und die zweite Auflage des Werkes sind bei Gondolat Könyvkiadó, Budapest erschienen

Gemeinschaftsausgabe des Verlag Dokumentation Saur KG, Pullach bei München, mit Akadémiai Kiadó, Budapest

Printed in Hungary · Satz und Druck: Akadémiai Nyomda, Budapest
ISBN 3-7940-4109-7

PREFACE TO THE SIXTH EDITION

In the last ten years, six editions of the Librarian's Practical Dictionary have been published for the use of library cataloging departments. The first and second editions (1963—1964) were geared toward librarians whose leading language was Hungarian, while the main language of the third, fourth, and fifth editions (1969—1970—1971) is German. The present sixth edition is based on English, since the practical significance of English in the world of library science is a long-standing fact.

The favorable reception of the previous editions has shown that my main purposes — easy handling, practicality and international use — have been achieved. This is why the editor has retained the same goals for this sixth, completely revised and enlarged edition, with English as the main language. These principles are the following:

1. The number of languages was dictated by actual needs. For this reason, two outstanding literary languages, Portuguese and Norwegian, have been added to the list, thus filling a long-felt need. Most reference works of a similar nature cover hardly more than ten languages, omitting those whose literature raises everyday problems for librarians. In this dictionary the source language is English, followed by French, German, Russian and Spanish. The remaining seventeen languages are listed in English alphabetical order, beginning with Bulgarian.

2. To insure practicality, a special layout was needed. My aim was to provide the librarian with a useful aid which would help in cataloging, arranging, and provide guidance in foreign documentation. The dictionary consists of two parts. Part I contains the word lists in tabular form; Part II gives the word lists in the individual languages successively. This dictionary is a combination of a general and special dictionary, in that it also contains some particles, articles, suffixes, prepositions, etc., that are not strictly relevant to librarianship or library science. These elements should aid users with little or no knowledge of the particular languages in making catalog entries. Calendar units for cataloging periodicals are given in special tables. There are also tables of the full forms of Arabic and Roman cardinal and ordinal numbers, as well as the international transliteration of Greek and Cyrillic letters. Four languages — Greek, Bulgarian, Russian and Serbian — are given both in the original Greek and Cyrillic script and in the latinized form, according to the rules of international transliteration. This is for the benefit of those not familiar with non-Roman alphabets, as well as for librarians working with entries transliterated into the Latin alphabet. Synonyms appear under the headwords (e.g. **drama** → **play**) with the usual cross-references.

3. The dictionary is suitable for international use. Knowledge of one language is sufficient, unless the user's only language is English. The basic terms are furnished with index numbers referring to the same number in all 22 languages. For example, the definite article "the" is listed in each language under number 339. A practical example: should a French librarian come across the unfamiliar Finnish word "kirjasto", he has to look up the Finnish word list (Part II), where he will find the index number 183 attached to the word "kirjasto"; then, on looking up the same number in the French column of Part I, there he will find the French equivalent term: "bibliothèque".

4. The 22 languages can be used in 462 different bilingual combinations, thus forming 462 bilingual word lists (glossaries). Including the corresponding synonyms, the total number of dictionary entries is approximately 17,000. On the other hand, nation names, which took up too much space in previous editions, were excluded from the present one. A dictionary of such specific nature cannot aim at completeness in every respect.

Since this is a very special dictionary, general dictionaries alone could not be relied upon. In addition to my own experience, I turned to the most competent authorities in each language area, most of whom are active in national or university libraries. Thanks to the international solidarity of librarians, I was able to complete the present work. I gratefully record my obligation to the institutions which have been of assistance in supervising, regularly or occasionally, the correctness of the special terms submitted to them in the course of the past ten years:

Athens, Ethnike Bibliotheke
Belgrade, Jugoslavenski Bibliografski Institut
Bratislava, Universitná Knižnica
Bucharest, Biblioteca Centrală de Stat
Copenhagen, Kongelige Bibliotek
The Hague, Koninklijke Bibliotheek
Helsinki, Yliopiston Kirjasto
Lisbon, Biblioteca Nacional
Madrid, Biblioteca Nacional
Moscow, Vsesoyuznaya Knizhnaya Palata
Oslo, Universitetsbibliotek
Paris, Bibliothèque Nationale
Prague, Universitni Knihovna
Rome, Biblioteca Nazionale Centrale
Sofia, Narodna Biblioteka
Stockholm, Kungliga Biblioteket
Warsaw, Biblioteka Narodowa
Zagreb, Nacionalna i Sveučilišna Biblioteka

Contacts with these renowned institutions were established by the Országos Széchényi Könyvtár, the National Library of Hungary (founded in 1802) through the good offices of its deputy director Dr. G. Sebestyén. The present, sixth edition of Dictionarium Bibliothecarii Practicum is the result of these fruitful relations.

Budapest, May 1973   *Dr. Z. Pipics*

VORWORT ZUR SECHSTEN AUSGABE

Das D i c t i o n a r i u m B i b l i o t h e c a r i i P r a c t i c u m erschien im Laufe der vergangenen zehn Jahre in sechs, voneinander abweichenden Ausgaben. Für die erste und zweite Auflage (1963 und 1964) war die ungarische Sprache die Ausgangssprache, für die dritte, vierte und fünfte Auflage (1969, 1970 und 1971) war es die deutsche, und jetzt, für die vorliegende, sechste Auflage bildet die englische Sprache die Ausgangssprache. Es erübrigt sich, die praktische Bedeutung dieser Weltsprache für die Bibliothekswissenschaft überhaupt zu beweisen. Die verschiedenen Ausgaben des Dictionarium sind ein wichtiges Hilfsmittel für die Katalogisierungsarbeiten der Bibliotheken, und findet sich in den Regalen aller Handbibliotheken.

Das positive Echo der früheren Auflagen dieses Wörterbuches bestätigte mir, daß es richtig war, vor allem auf folgende Hauptpunkte bei der Bearbeitung zu achten: den der praktischen Anwendungsmöglichkeit und den der Eignung zum internationalen Gebrauch. Demnach habe ich in der vorliegenden sechsten, neubearbeiteten und erweiterten Ausgabe — diesmal mit Englisch als Grundsprache — die ursprünglichen Prinzipien beibehalten:

1. Die Anzahl der Sprachen richtet sich nach dem tatsächlichen Bedarf. Es wurden deshalb zwei weitere Sprachen: das N o r w e g i s c h e und das P o r t u g i e s i s c h e — beide mit bedeutender Literatur — aufgenommen, wodurch ein bereits seit langem verspürter Mangel behoben wurde. Die meisten ähnlichen bibliothekarischen Werke erfassen kaum mehr als zehn Sprachen und lassen solche außer acht, deren Literatur für die Bibliothekare ein tägliches Problem bedeutet. In diesem Wörterbuch ist die Grundsprache das Englische, dem Französische, Deutsche, Russische und Spanische folgen, sodann kommen mit dem Bulgarischen an der Spitze weitere siebzehn Sprachen, in der Reihenfolge des englischen Alphabets.

2. Die praktische Bestimmung des Wörterbuches erforderte einen speziellen Aufbau. Das Ziel war, dem Bibliothekar zu seiner alltäglichen Arbeit auf dem Gebiete der Katalogisierung und Ordnung fremdsprachiger Dokumente eine Hilfe zu bieten. Das Wörterbuch hat zwei Teile: I. Das in Tabellen gefaßte gemeinsame Wörterverzeichnis der Sprachen; II. Die Wörterverzeichnisse der einzelnen Sprachen. Das Wörterbuch stellt eine Kombination der allgemeinen und der Fachwörterbücher dar, insofern es auch zahlreiche grammatikalische Hilfswörter wie Artikel, Suffixe, Präpositionen usw. enthält, die nicht unmittelbar in den Bereich des Bibliothekswesens und der Bibliothekswissenschaft gehören. Diese erleichtern die Konstruierung der Titelaufnahmen für diejenigen, die die betreffende Sprache überhaupt nicht oder nur in geringem Maße kennen. Spezielle Tabellen enthalten die Benennungen der zur Bearbeitung der Periodika benötigten Zeitabschnitte, die Auflösung der arabischen und römischen Grund- und Ordnungszahlwörter in Buchstaben sowie die bibliothekarischen internationalen Transkriptionsregeln der griechischen und kyrillischen Buchstaben. Vier Sprachen: Griechisch, Bulgarisch, Russisch und Serbisch sind außer der ursprünglichen griechischen bzw. kyrillischen Schrift auch in der durch die internationale Norm angenommenen Transkription in lateinischer Schrift gebracht, für Anfänger und diejenigen Bibliothekare, die diese Schriften nicht kennen bzw. die mit transkribierten Titelaufnahmen arbeiten. Die sinnverwandten Wörter werden unter einem Sammelwort zusammengefaßt (z. B. **drama** → **play**).

3. Das Wörterbuch ist für den internationalen Gebrauch geeignet. Von den zweiundzwanzig Sprachen genügt bereits die Kenntnis einer einzigen, falls das Wörterbuch nicht von einem englischen Bibliothekar benutzt werden sollte. Die Grundbegriffe sind nämlich mit Leitzahlen versehen, die sich auf dieselben Zahlen in allen zweiundzwanzig Sprachen beziehen. So steht z. B. der bestimmte Artikel (the) im Wörterverzeichnis einer jeden Sprache unter der Leitzahl 339. An einem praktischen Beispiel erläutert: begegnet einem französischen Bibliothekar das finnische Wort 'kirjasto', das er nicht kennt, so sucht er die im finnischen Wörterverzeichnis (Teil II) bei dem Wort 'kirjasto' die an dem Seitenrand stehende Leitzahl 183 in der französischen Kolumne des Gesamtregisters (Teil I) heraus, wo er das Wort 'bibliothèque' und damit auch die Lösung finden wird.

4. Die Wörter aus zweiundzwanzig Sprachen lassen sich 462fach einander gegenüberstellen, und damit erübrigen sich 462 zweisprachige Wörterverzeichnisse (Glossarien). Mit den entsprechenden Synonymen wächst die Anzahl der Ausdrücke auf etwa 17 000 lexikographische Angaben an. Hingegen wurden die Völkerbezeichnungen (die in den vorangehenden Auflagen allzu viel Raum eingenommen haben) außer acht gelassen, in der Überzeugung, daß ein Wörterbuch spezieller Art keinen Anspruch auf Vollständigkeit erheben kann.

Da es sich um ein Fachwörterbuch ganz eigener Art handelt, kamen allgemeine Sprachwörterbücher als etwaige Quellenwerke nicht in Betracht. Wo meine eigenen Kenntnisse und Erfahrungen nicht ausreichten, mußte ich mich an die authentischen Kenner der Fachsprache der Großbibliotheken in anderen Ländern wenden. Für die internationale Solidarität der Bibliothekare sollen hier mit verbindlichem Dank die Namen all derjenigen Institutionen aufgezählt werden, die in den vergangenen zehn Jahren regelmäßig oder fallweise die Authentizität der Fachausdrücke geprüft haben:

```
Athen, Ethnike Bibliotheke
Belgrad, Jugoslavenski Bibliografski Institut
Bratislava, Universitná Knižnica
Bukarest, Biblioteca Centrală de Stat
Den Haag, Koninklijke Bibliotheek
Helsinki, Yliopiston Kirjasto
Kopenhagen, Kongelige Bibliotek
Lissabon, Biblioteca Nacional
Madrid, Biblioteca Nacional
Moskau, Wsesojusnaja Knishnaja Palata
Oslo, Universitetsbibliotek
Paris, Bibliothèque Nationale
Prag, Universitni Knihovna
Rom, Biblioteca Nazionale Centrale
Sofia, Narodna Biblioteka
Stockholm, Kungliga Biblioteket
```

Warschau, Biblioteka Narodowa
Zagreb, Nacionalna i Sveučilišna Biblioteka

Die Beziehungen zu diesen hochangesehenen Institutionen hat die ungarische Nationalbibliothek Országos Széchényi Könyvtár (gegründet 1802) — persönlich ihr Stellvertretender Oberdirektor Dr. G. Sebestyén — von Anfang an mit großem Verständnis ausgebaut. Das Ergebnis dieser ersprießlichen Tätigkeit bildet die vorliegende 6. Ausgabe.

Budapest, Mai 1973

Dr. Z. Pipics

AD USUM BIBLIOTHECARIORUM NON-ANGLICORUM

Exquirendo traductionem cuiuslibet termini in aliam linguam, consultamus primum partem secundam vocabularii (Linguae singulae). Numeri indices inveniuntur in margine.

Dein inquirendus est numerus necessarius in rubrica conveniente partis primae vocabularii (Linguae cunctae).

A L'ATTENTION DES BIBLIOTHÉCAIRES NON ANGLAIS

Pour trouver la traduction du terme inconnu dans la langue demandée, consulter d'abord Part II du vocabulaire (Linguae singulae), en cherchant le terme dans le registre alphabétique de la langue à laquelle ce term appartient. Noter la référence numérique de renvoi portée en marge.

Identifier l'équivalent du terme en question dans le tableau synoptique de la Part I du vocabulaire (Linguae cunctae), en cherchant, d'après la référence numérique de renvoi, la rubrique de la langue intéressée.

ДЛЯ БИБЛИОТЕКАРЕЙ, НЕ ПОЛЬЗУЮЩИХСЯ АНГЛИЙСКИМ ЯЗЫКОМ

При каталогизации и переводе терминов в первую очередь следует пользоваться второй частью словаря (Linguae singulae). Индекс-номер на поле страницы.

После этого следует разыскать индекс-номер в соответствующей рубрике первой части словаря (Linguae cunctae).

₫ARA LOS QUE EMPLEAN ESTE DICCIONARIO PERO NO CONOCEN EL IDIOMA INGLES

Queriendo saber el significado de un vocablo en otro idioma, hay que recurrir al Parte II (Linguae singulae), donde el indice en el margen del vocabulario servirá para orientarse.

Con la ayuda de este índice se encontrará en la Parte I (Linguae cunctae) la acepción del vocablo en los otros idiomas.

ABBREVIATIONS—(Notae)—ABKÜRZUNGEN—ABBRÉVIATIONS—СОКРАЩЕНИЯ—ABBREVIATURAS

a.   adjective — (adjectivum) — adjectif — Adjektiv — прилагательное — adjetivo
n.   noun — (substantivum) — substantiv, nom — Substantiv — существительное -- substantivo
pl.   plural — (pluralis) — pluriel — Plural — множественное число — plural
v.   verb — (verbum) — verbe — Verb — глагол — verbo
→   see — (vide) — voir — siehe — смотри — véase
/   or — (sive) — ou — oder — или — o
—   omitted in translation — (in alium sermonem non convertitur) — ne pas traduire — bleibt unübersetzt — не переводится — no se traduce

11

I

	ENGLISH Anglica	FRENCH Gallica	GERMAN Germanica	RUSSIAN Russica	SPANISH Hispanica
1	**a, an**	un, une	ein, eine, ein	odin, odna, odno (один, одна, одно)	un(o), una
2	**abbreviated entry**	notice abrégée	verkürzte Aufnahme	sokraščёnnoe/ kratkoe opisanie (сокращённое/ краткое описание)	asiento bibliográfico abreviado
3	**about**	de, d' sur	über um	o (o) ob (об) obo (обо)	de
4	**abridge**	abréger	abkürzen	sokraščať (сокращать)	abreviar acortar
	abstract →**summary**				
5	**academy**	académie	Akademie	akademija (академия)	academia
	account →**report**[1]				
6	**act**	acte	Aufzug Akt	akt (акт) dejstvie (действие)	acto jornada
7	**adage** **(aphorisme** **saying)**	adage dicton	Spruch	pogovorka (поговор- ка) izrečenie (изречение) aforizm (афоризм) maksima (максима)	sentencia máxima
8	**adapt**	adapter	bearbeiten	adaptirovať (адаптировать)	adaptar aplicar
9	**added entry** **(secondary entry)**	entrée secondaire	Nebeneintragung	dobavočnoe opisanie (добавочное описание) vspomogateľnoe opisanie (вспомогательное описание)	encabezamiento secundario (España) entrada secundaria (Hispanoamérica) ficha secundaria

BULGARIAN Bulgarica	CROATIAN Croatica	CZECH Bohemica	DANISH Danica	DUTCH Hollandica	FINNISH Fennica
edin, edna, edno (един, една, едно)	jedan, jedna, jedno	jeden, jedna, jedno	en et	een	—
sâkrateno opisanie (съкратено описание)	skraćeni opis	zkrácený popis	forkórtet indførsel	verkorte titelbeschrijving verkorte catalogustitel	lyhennetty kirjaus
otnosno (относно) vârhu (върху) za (за) po (по)	o po vrh	o	om over	over van	-sta -stä
sâkraštavam (съкращавам) skâsjavam (скъсявам)	skraćivati	zkrátit	forkorte	afkorten verkorten	lyhentää
akademija (академия)	akademija	akademie	akademi	academie	akatemia
dejstvie (действие)	čin akt	dějství jednání	akt	akte bedrijf	näytös
pogovorka (поговорка) maksima (максима)	mudra izreka	průpověď výrok	sentens	spreekwoord	mietelause mietelmä
adaptiram (адаптирам)	adaptirati prilagoditi	přizpůsobit upravit	tillempe	aanpassen aanwenden	laittaa sovittaa
dopâlnitelno opisanie (допълнително описание)	sporedna kataloška jedinica	vedlejši záznam	bikort	opneming onder secundaire hoofd- woord	lisäkirjaus apukirjaus apukortti

15

	GREEK Graeca	HUNGARIAN Hungarica	ITALIAN Italica	LATIN Latina	NORWEGIAN Norvegica
1	heis *(εἷς)* hen *(ἕν)* mia *(μία)* ti(s) *(τὶς)*	egy —	un, uno una un'	unus, una, unum	en, et en, ett
2	syntomos anagraphe *(σύντομος ἀναγραφή)*	rövidített címfelvétel	indicazione abbreviata	catalogisatio brevis	forkortet innførsel
3	gia *(γιά)* dia *(διά)* peri *(περί)*	-ról, -ről; felett	da su sopra	de supra super	om over
4	brachyno *(βραχύνω)* systello *(συστέλλω)*	rövidít	abbreviare compendiare riassumere	breviorem facere compendere contrahere	forkorte
5	akademeia *(ἀκαδήμεια)*	akadémia	accademia	academia	akademi
6	praxis *(πρᾶξις)*	felvonás	atto	actus	akt
7	aphorisma *(ἀφόρισμα)* apophthegma *(ἀπόφθεγμα)* gnome *(γνώμη)*	mondás	detto sentenza	adagium dictum maxima sententia verbum	ordspråk sentens
8	enarmozo *(ἐναρμόζω)*	alkalmaz felhasznál	adattare applicare	accommodare adaptare aptare	avpasse bearbeide tillempe
9	sympleromatikon deltion *(συμπληρωματικόν δελτίον)*	melléklap	scheda secondaria	charta/scida adiuncta charta/scida secunda	tilleggsinnførsel biinnførsel

POLISH Pol	PORTUGUESE Portugallica	RUMANIAN Rumenica	SERBIAN Servica	SLOVAK Slovaca	SWEDISH Suecica
jeden, jedna, jedno	um(a)	un, o; nişte *pl.*	jedan, jedna, jedno (један, једна, једно)	jeden, jedna, jedno	en ett
skrócony opis bibliograficzny	entrada simplificada	catalogizaţie scurtată	skraćeni opis (скраћени опис)	skrátený popis	förkortad titel
o nad	acerca de de sobre, sôbre	asupra despre	o (о) po (по) vrh (врх)	o	om över
skracać	abreviar resumir	abrevia prescurta	skraćivati (скраћивати)	krátiť	förkorta
akademia	academia	academie	akademija (академија)	akadémia	akademi
akt	acto	act	čin (чин) akt (акт)	dejstvo	akt
aforyzm przypowieść sentencja	adagio provérbio	maximă sentinţă vorbă	(mudra) izreka (мудра изрека)	maxima výrok	ordspråk sentens
adaptować przystosować	adaptar aplicar	adapta aplica	adaptovati (адаптовати) prilagoditi (прилагодити)	prispôsobiť upraviť	anpassa avpassa
opis dodatkowy	entrada secundária	fişă suplimentară/ secundară	sporedna kataloška jedinica (споредна каталошка јединица)	vedľajší záznam	biuppslag

	ENGLISH Anglica	FRENCH Gallica	GERMAN Germanica	RUSSIAN Russica	SPANISH Hispanica
10	**addenda** *pl.*	addenda *pl.*	Nachträge *pl.* Ergänzungen *pl.*	dopolnenie (дополнение) dobavlenie (добавление)	complemento suplemento
11	**advertisement**	annonce publicitaire	Reklame Inserat	ob''javlenie (объявление) reklama (реклама)	anuncio
12	**after**	après	nach	posle (после)	detrás de después de
13	**album**	album	Album	aľbom (альбом)	álbum
14	**almanac(k)**	almanach	Almanach	aľmanach (альманах)	almanaque

15	**alphabetical order**	classement
		  alphabétique
			alphabetische
			  Ordnung
				alfavitnyj
				  porjadok
				(алфавитный
				  порядок)
					orden
					  alfabético

16	**alphabetize**	ranger par ordre
		  alphabétique
			alphabetisieren	raspolagať v
				alfavitnom
				  porjadke
				(располагать в
				алфавитном
				  порядке)
					alfabetizar

	alter →**change**[1]				
17	**ancient** **(old)**	ancien vieux antique	alt	staryj (старый) drevnij (древний)	antiguo viejo
18	**and**	et	und	i (и)	y
19	**annals** *pl.*	annales *pl.*	Jahrbücher *pl.* Annalen *pl.*	ežegodnik (ежегодник)	anuario anales *pl.*
20	**annex**	joindre inclure	beilegen beischließen	prikladyvať (прикладывать) prilagať (прилагать)	adjuntar incluir

BULGARIAN Bulgarica	CROATIAN Croatica	CZECH Bohemica	DANISH Danica	DUTCH Hollandica	FINNISH Fennica
dobavki (добавки) *pl.* dopâlnenie (допълнение)	dopuna dodatak	dodatek	supplement tillæg	supplement aanvulling	täydennys
objavlenie (обявление) reklama (реклама)	oglas obavijest reklama	inzerát reklama	annonce avertissement	advertentie reclame	ilmoitus mainos
sled (след) po (по)	poslije iza	po podle	efter	naar	jälkeen
album (албум)	album	album	album	album	albumi
almanah (алманах)	almanah	almanach	almanak	almanak	almanakka
azbučen red (азбучен ред)	abecedni red	abecední pořadí	alfabetisk ordning	alfabetische rangschikking	aakkosjärjestys
nareždam po azbučen red (нареждам по азбучен ред)	slagati po abecedi abecedirati	srovnat abecedně	alfabetisere	alfabetiseren	aakkostaa
star (стар) dreven (древен)	star antički	starý dávný antický	gammel	oud	vanha entinen muinainen
i (и)	a i pa	a i	og	en	ja
letopisi (летописи) *pl.*	ljetopisi *pl.*	anály *pl.* letopisy *pl.*	annaler *pl.*	annalen *pl.*	vuosikirjat *pl.* annaalit *pl.*
pribavjam (прибавям)	prilagati pridodavati	připojit	føje til	bijdragen bijvoegen	liittää lisätä

19

		GREEK Graeca	HUNGARIAN Hungarica	ITALIAN Italica	LATIN Latina	NORWEGIAN Norvegica
10		prosthekai (προσθήκαι) pl.	kiegészítés(ek pl.)	supplemento aggiunta	supplementum complementum	bilag supplement tillegg
11		angelia (ἀγγελία) diaphemisis (διαφήμισις)	reklám hirdetés	pubblicità	propaganda	annonse avertissement
12		meta (μετά) kata (κατά)	után	dopo	post secundum, ad	etter
13		leukoma (λεύκωμα) almpum (ἄλμπουμ)	album	albo album	album	album
14		almanak (ἀλμανάκ) almanach (ἀλμανάχ)	almanach	almanacco	almanachus	almanakk
15		alphabetike taxino- mesis (ἀλφαβητική ταξινόμησις)	betűrend abécérend	ordinamento alfabetico	litterarum ordo alphabetum	alfabetisk ordning
16		aradiazo ste alphabetike seira (ἀραδιάζω στή ἀλφαβητική σειρά)	betűrendez ábécéz	alfabetare	in litteram digerere alphabeticare	alfabetisere
17		pal(a)ios (παλαιός) archaios (ἀρχαῖος)	régi ó	antico vecchio	antiquus vetus	gammel
18		kai (καί)	és, s	e ed	et	og
19		chronika (χρονικά) pl.	évkönyvek pl.	annali pl.	annales pl.	annaler pl.
20		prostithemi (προστίθημι) esokleio (ἐσωκλείω)	csatol mellékel	aggiungere annettere	adiungere annectere addere	sette til

POLISH Polonica	PORTUGUESE Portugallica	RUMANIAN Rumenica	SERBIAN Servica	SLOVAK Slovaca	SWEDISH Suecica
dopełnienie dodatek	complemento suplemento	supliment	dopuna (допуна) dodatak (додатак)	doplnok	tillägg supplement
reklama	anúncio	reclamă publicaţie	reklama (реклама)	inzerát reklama	annons
po za	depois de	după	posle (после) iza (иза)	po za	efter
album	álbum	album	album (албум)	album	album
almanach	almanaque	almanah	almanah (алманах)	almanach	almanack(a)
porządek abecadłowy	alfabetacão ordem alfabetica	ordine alfabetică	azbučni/ alfabetski red (азбучни/ алфабетски ред)	abecedný poriadok	alfabetisk ordning
układać podług alfabetu	alfabetar	alfabetiza	slagati po abecedi (слагати по абецеди)	usporiadať abecedne	alfabetisera
dawny	antigo velho	antic vechi	star (стар) antički (антички)	starý dávny antický	gammal
i	e	şi	a (а) i (и) pa (па)	a i	och
annały *pl.*	anais *pl.*	anuare *pl.* anale *pl.*	letopisi (летописи) *pl.* anali (анали) *pl.*	anály *pl.* letopisy *pl.*	annaler *pl.*
dołączyć załączyć	juntar	adăuga la anexa la ataşa la	prilagati (прилагати) pridodavati (придодавати)	priložiť pripäť pripojiť	bifoga tillägga

		ENGLISH Anglica	FRENCH Gallica	GERMAN Germanica	RUSSIAN Russica	SPANISH Hispanica
21		**annotate**	annoter	annotieren	annotirovať (аннотировать)	anotar
22		**annotation** **(explanatory note note)**	annotation	Annotation Anmerkung	annotacija (аннотация) primečanie (примечание)	anotación nota
		annual **→year-book**				
		annual volume **→volume²**				
23		**anonymous**	anonyme	anonym verfasserlos	anonimnyj (анонимный)	anónimo
24		**anthem** **(hymn)**	hymne	Hymne	gimn (гимн)	himno
25		**anthology** **aphorism→adage**	anthologie	Anthologie	antologija (антология)	antología
		appear→ **be published**				
26		**appendix**	appendice	Anhang Zusatz Appendix	priloženie (приложение)	apéndice
27		**applied**	appliqué	angewandt	prikladnoj (прикладной)	aplicado
28		**archives** *pl.* **(record office)**	dépôt des actes archives *pl.*	Archiv	archiv (архив) archivnoe učreždenie (архивное учреждение)	archivo
		argument→contents				
29		**arrange** **(order³)**	arranger ordonner	ordnen	raspolagať (располагать) privodiť v porjadok (приводить в порядок)	arreglar ordenar

BULGARIAN Bulgarica	CROATIAN Croatica	CZECH Bohemica	DANISH Danica	DUTCH Hollandica	FINNISH Fennica
anotiram (анотирам)	pribilježiti anotirati	anotovat	annotere forsyne med noter	annoteren	esitellä
anotacija (анотация)	bilješka napomena primjedba opaska	anotace poznámka	anmærkning note	aantekening annotatie noot	selostus
anonimen (анонимен) bezimen(en) (безименен)	anoniman bezimen	anonymní bez autora	anonymní	anoniem naamloos	anonyymi nimetön
himn (химн)	himna	hymna	hymne	hymne	hymni
antologija (антология)	antologija	antologie	antologi	anthologie	runokokoelma runovalikoima
priloženie (приложение) priturka (притурка) pribavka (прибавка)	dodatak dopuna(k)	dodatek přídavek	tillæg appendiks supplement	aanhangsel	liite lisä
priložen (приложен)	primijenjen	užitý	anvendt	toegepast	sovellettu
arhiv (архив) arhiva (архива)	arhiv	archív	arkiv	archief	arkisto
podreždam (подреждам) privezdam v red (привеждам в ред)	prirediti	uspořádat	arrangere ordne	ordenen rangschikken	järjestää panna kuntoon

	GREEK Graeca	HUNGARIAN Hungarica	ITALIAN Italica	LATIN Latina	NORWEGIAN Norvegica
21	semeiono (σημειώνω)	annotál	annotare	annotare	annotere forsyne med anmerkninger
22	semeiosis (σημείωσις)	jegyzet	annotazione nota	nota glossa annotatio	anmerkning note
23	anonymos (ἀνώνυμος)	névtelen	anonimo	anonymus	anonym
24	hymnos (ὕμνος)	himnusz	inno	hymnus	hymne
25	anthologia (ἀνθολογία)	antológia	antologia	florilegium anthologia	antologi
26	parartema (παράρτημα) parergon (πάρεργον) prostheke (προσθήκη) prosthesis (πρόσθεσις)	függelék toldalék pótlás	appendice	appendix additamentum	appendiks bilag tillegg
27	ephermosmenos (ἐφηρμοσμένος)	alkalmazott	applicato	applicatus	anvendt
28	archeion (ἀρχεῖον)	levéltár archivum	archivio	archivum chartarium	arkiv
29	taktopoio (τακτοποιῶ) dioikeo (διοικέω)	rendez	assortire ordinare	ordinare componere digerere	opstille ordne

POLISH Polonica	PORTUGUESE Portugallica	RUMANIAN Rumenica	SERBIAN Servica	SLOVAK Slovaca	SWEDISH Suecica
adnotować	anotar	adnota	anotirati (анотирати) pribeležiti (прибележити)	anotovať	annotera förse med noter
adnotacja dopisek notatka	anotação	însemnare notă	opaska (опаска beleška (белешка) napomena (напомена) primedba (примедба)	anotácia poznámka	anmärkning not
anonimowy	anónimo	anonim	anoniman (анониман) bezimen (безимен)	anonymný	anonym
hymn	hino	imn	himna (химна)	hymna	hymn
antologia	antologia	antologie	antologija (антологија)	antológia	antologi
dodatek suplement uzupełnienie	apêndice suplemento	apendice	dodatak (додатак) dopuna (допуна)	dodatok	tillägg bihang
stosowany	aplicato	aplicat	primenjen (примењен)	užitý	tillämpad
archiwum	arquivos *pl.*	arhivă	arhiv (архив)	archív	arkiv
porządkować	ordenar pôr em ordem	rîndui aranja aşeza	prirediti (приредити)	usporiadať upraviť	ordna

		ENGLISH Anglica	FRENCH Gallica	GERMAN Germanica	RUSSIAN Russica	SPANISH Hispanica
30		art	art	Kunst	iskusstvo (искусство)	arte
31		article	article	Artikel	statja (статья)	artículo
32		assembly (meeting)	assemblée meeting	Versammlung Tagung	sobranie (собрание)	asamblea junta
		assist →collaborate				
33		atlas	atlas	Atlas	atlas (атлас)	atlas
34		authentic	authentique	authentisch echt beglaubigt	autentičeskij (аутентический) podlinnyj (подлинный)	auténtico
35		author	auteur	Verfasser	avtor (автор)	autor
36		author bibliography (biobibliography)	bibliographie individuelle biobibliographie	Personalbiblio- graphie Biobibliographie	personalnaja bibliografija (персональная библиография) biobibliografija (биобиблиография)	bibliografía personal bio-bibliografía
37		authorize	autoriser	autorisieren berechtigen	avtorizovať (авторизовать)	autorizar
38		autobiography	autobiographie	Selbstbiographie Autobiographie	avtobiografija (автобиография)	autobiografía
39		autograph	manuscrit autographique autographe	Autograph	avtograf (автограф) original rukopisi (оригинал рукописи)	manuscrito autográfico autógrafo
40		back (spine)	dos	Rücken	korešok (корешок)	lomo
41		based/founded on the . . . (on the basis of . . .)	fondé sur . . .	auf Grund des . . .	na osnovanii (на основании . . .)	sobre la base . . .

BULGARIAN Bulgarica	CROATIAN Croatica	CZECH Bohemica	DANISH Danica	DUTCH Hollandica	FINNISH Fennica
izkustvo (изкуство)	umjetnost	umění	kunst	kunst	taide
člen (член) statija (статия)	članak	článek pojednání	artikel	artikel	artikkeli kirjoitus
sâbranie (събрание) sbirka (сбирка)	sastanak zbor	shromáždění sjezd	forsamling møde	vergadering bijeenkomst	kokous
atlas (атлас)	atlas	atlas	atlas	atlas	atlas kartasto
avtentičen (автентичен) istinski (истински)	autentičan pravi	autentický ověřený	ægte autentisk stadfæstet	authentiek echt	autenttinen oikeaperäinen
avtor (автор)	autor	autor původce	forfatter	auteur schrijver	tekijä
personalna bibliografija (персонална библиография) biobibliografija (биобиблиография)	osobna bibliografija	personální bibliografie	personalbibliografi	personeele bibliografie	henkilöbibliografia
avtoriziram (авторизирам)	autorizirati ovlastiti	autorizovat oprávnit zmocnit	bemyndige berettige	machtigen	oikeuttaa valtuuttaa
avtobiografija (автобиография)	autobiografija	autobiografie	selvbiografi	autobiografie	omaelämäkerta autobiografia
avtograf (автограф)	izvorni rukopis	originál autograf	originalhåndskrift	autograaf	alkuperäinen käsikirjoitus
grâb na kniga (гръб на книга)	hrbat	hřbet knihy	ryg	rug	selkä
vâz osnova na (въз основа на . . .)	na osnovu . . . na temelju . . .	na podkladě . . .	på grundlag af . . .	op grond van . . .	perusteella pohjalla

		GREEK Graeca	HUNGARIAN Hungarica	ITALIAN Italica	LATIN Latina	NORWEGIAN Norvegica
30		techne *(τέχνη)*	művészet	arte	ars	kunst
31		arthron *(ἄρθρον)*	cikk(ely) dolgozat, cikk	articolo	articulus elucubratio	artikkel
32		syneleusis *(συνέλευσις)* symposion *(σημπόσιον)*	gyűlés	adunanza convegno	concilium conferentia congressus conventus	forsamling møte
33		atlas *(ἄτλας)*	atlasz térképgyűjtemény	atlante	atlas	atlas
34		authentikos *(αὐθεντικός)*	hiteles	autentico	authenticus verus	autentisk ekte
35		syngrapheus *(συγγραφεύς)*	szerző	autore	auctor conditor scriptor	forfatter
36		atomike bibliographia *(ἀτομική βιβλιο- γραφία)*	szerzői bibliográfia	bibliografia personale	bibliographia auctoris/individualis	personalbibliografi
37		epitrepo *(ἐπιτρέπω)* exusian parecho *(ἐξουσίαν παρέχω)*	jogosít	autorizzare	auctorisare	autorisere bemyndige berettige
38		autobiographia *(αὐτοβιογραφία)*	önéletrajz	autobiografia	autobiographia	selvbiografi
39		autographon *(αὐτόγραφον)* cheirographon *(χειρόγραφον)*	eredeti kézirat	autografo	autographum	originalhåndskript originalmanuskript
40		rahis *(ῥάχις)*	gerinc könyvgerinc	dorso	dorsum libri	rygg
41		(epi te) basei ... *(ἐπί τῇ βάσει ...)*	... alapján	in base a ...	in base ... ex...	på grunn av ...

POLISH Polonica	PORTUGUESE Portugallica	RUMANIAN Rumenica	SERBIAN Servica	SLOVAK Slovaca	SWEDISH Suecica
sztuka	arte	artă	umetnost (уметност)	umenie	konst
artyikuł	artigo	articol	članak (чланак)	článok	artikel
zebranie zgromadzenie	assembleia reunião	adunare întrunire	sastanak (састанак) zbor (збор)	zhromaždenie zjazd	församling möte sammankomst
atlas	atlas	atlas	atlas (атлас)	atlas	atlas kartbok
autentyczny	autêntico	autentic	autentičan (аутентичан) pravi (прави)	autentický hodnoverný	autentisk trovärdig
autor	autor	autor	autor (аутор)	autor	författare
bibliografia osobo- wa	bibliografia pessoal	bibliografie de autor	personalna bibliografija (персонална библиографија)	personálna bibliografia	personbibliografi biobibliografi
autoryzować upoważnić	autorizar	autoriza	autorizovati (ауторизовати) ovlastiti (овластити) odobriti (одобрити)	autorizovať oprávňovať splnomocniť	auktorisera
autobiografia	autobiografia	autobiografie	autobiografija (аутобиографија)	autobiografia vlastný životopis	självbiografi
autograf	autógrafo	autograf	originalni rukopis (оригинални рукопис)	autografický ruko- pis	autograf
grzbiet	lombada	dos	hrbat (хрбат)	chrbát knihy	rygg
na podstawie . . .	na/com base de . . .	bazat pe . . . pe/în baza . . . pe bază de . . .	na osnovu . . . (на основу . . .)	na základe . . . podľa . . .	på grund av . . .

29

		ENGLISH Anglica	FRENCH Gallica	GERMAN Germanica	RUSSIAN Russica	SPANISH Hispanica
42	bibliography	bibliographie	Bibliographie Literaturverzeichnis Bücherkunde	bibliografija (библиография)	bibliografía
43	bibliology	bibliologie	Buchwesen Buchkunde	knigovedenie (книговедение)	bibliología
	big→large				
44	bilingual	bilingue	zweisprachig	dvujazyčnyj (двуязычный)	bilingüe
45	bind	relier	binden	perepletať (переплетать)	encuadernar

46	bindery	atelier de reliure	Buchbinderei	pereplëtnaja
				(переплётная)
					taller de
					  encuadernador/
					  /encuadernación

47	binding	reliure	Einband	pereplët (переплёт)	encuadernación
	biobibliography →author bibliography				
48	biography	biographie	Biographie Lebensbeschreibung	biografija (биография) žizneopisanie (жизнеописание)	biografía
	body size →type size				
49	book	livre	Buch	kniga (книга)	libro
50	booklet	brochure	Heft Broschüre	brošjura (брошюра) tetrađ (тетрадь)	cuaderno
51	bookshop	librairie	Buchhandlung Buchladen	knižnyj magazin (книжный магазин)	librería
	bookstacks →stack-room				
	borough library →city library				

BULGARIAN Bulgarica	CROATIAN Croatica	CZECH Bohemica	DANISH Danica	DUTCH Hollandica	FINNISH Fennica
bibliografija (библиография)	bibliografija	bibliografie	bibliografi	bibliografie boekbeschrijving	bibliografia kirjatiede
knigoznanie (книгознание)	bibliologija	knihověda	bogkundskab	bibliologie	kirjatieto(us)
dvuezičen (двуезичен)	dvojezičan	dvoujazyčný	tosproget	tweetalig	kaksikielinen
podvârzvam (подвързвам)	uvezati ukoričiti	spojit vázat	binde indbinde	binden	sitoa
knigoveznica (книговезница)	knjigovežnica	knihařská dílna	bogbinderi	boekbinderij	kirjansitomo
podvârzija (подвързия)	uvez	vazba	inbinding	boekband	kirjansidos
biografija (биография) životopis (животопис)	biografija životopis	biografie	biografi levnedsbeskrivelse	biografie levensbeschrijving	elämäkerta
kniga (книга)	knjiga	kniha	bog	boek	kirja
brošura (брошура) knižka (книжка)	brošura sveska	brožura sešit	brochure hæfte	brochure heft	vihko
knižarnica (книжарница)	knjižara	knihkupectví	boghandel	boekhandel	kirjakauppa

	GREEK Graeca	HUNGARIAN Hungarica	ITALIAN Italica	LATIN Latina	NORWEGIAN Norvegica
42	bibliographia (βιβλιογραφία)	bibliográfia	bibliografia	bibliographia	bibliografi
43	bibliologia (βιβλιολογία)	könyvtudomány	bibliologia	bibliologia	bokkunskap
44	diglossos (δίγλοσσος)	kétnyelvű	bilingue	bilinguis	bilingval tospråklig
45	bibliodeto (βιβλιοδετῶ) deno (δένω)	köt	legare rilegare	alligare ligare	binde innbinde
46	bibliodeteion (βιβλιοδετεῖον)	könyvkötészet	legatoria	officina bibliopegi/ /glutinatoris	bokbinderi
47	bibliodesia (βιβλιοδέσια)	kötés bekötés	rilegatura	tegumentum/tegi- mentum libri	bind innbindning
48	biographia (βιογραφία)	életrajz	biografia	biographia	biografi
49	biblion (βιβλίον) biblos (βίβλος)	könyv	libro	liber libellus charta	bok
50	mprosura (μπροσούρα) phylladion (φυλλάδιον)	brosura füzet	quaderno	libellus plagulae iunctae	hefte
51	bibliopoleion (βιβλιοπωλεῖον)	könyvkereskedés könyvesbolt	libreria	taberna libraria	bokhandel

POLISH Polonica	PORTUGUESE Portugallica	RUMANIAN Rumenica	SERBIAN Servica	SLOVAK Slovaca	SWEDISH Suecica
bibliografia	bibliografia	bibliografie	bibliografija (библиографија)	bibliografia	bibliografi
nauka o książce	bibliologia	bibliologie	bibliologija (библиологија)	knihoveda	bokkunskap
dwujęzyczny	bilingue	bilingv	dvojezičan (двојезичан)	dvojjazyčný	tvåspråkig
wiązać	encadernar	lega	povezati (повезати) koričiti (коричити)	viazať	binda
introligatornia	encadernação	legătorie	knjigoveznica (књиговезница)	knihárska dielňa	bokbinderi
oprawa	encadernação	legătură	povez (повез)	väzba	band
biografia życiorys	biografia	biografie	biografija (биографија) životopis (животопис)	biografia životopis	biografi
książka księga	livro	carte	knjiga (књига)	kniha	bok
broszura zeszyt	brochura caderno	broşură cărţulie	brošura (брошура) sveska (свеска)	brožúra zošit	broschyr häfte
księgarnia	livraria	librărie	knjižara (књижара)	kníhkupectvo	bokhandel

	ENGLISH Anglica	**FRENCH** Gallica	**GERMAN** Germanica	**RUSSIAN** Russica	**SPANISH** Hispanica
52	braille printing (writing for the blind)	impression en braille	Blindenschrift	pis'mo/šrift dlja slepych (письмо/шрифт для слепых)	escritura Braille
53	bridal song (nuptial song)	chant nuptial	Hochzeitslied	svadebnaja pesnja (свадебная песня) ėpitalama (эпиталама)	epitalamio
	broadside →flysheet				
54	bulletin	bulletin	Bulletin Bericht	bjulleten' (бюллетень)	boletín
55	by	par	von	—	por
	call number →location mark				
56	card	fiche	Zettel, Karte Katalogzettel Katalogkarte Karteizettel Karteikarte	kartočka (карточка)	ficha
	card catalogue →card index				
57	card index (card catalogue)	catalogue sur fiches	Kartei Zettelkatalog	kartočnyj katalog/ ukazateľ (карточный каталог/ указатель)	catálogo en fichas
58	catalog(ue) *n.*	catalogue	Katalog	katalog (каталог)	catálogo
59	catalog(ue) *v.*	cataloguer	katalogisieren verzetteln	katalogizirovať (каталогизировать)	catalogar
60	cataloging (cataloguing entry description)	catalogage cataloguement	Titelaufnahme Katalogisierung	katalogizacija (каталогизация)	catalogación

BULGARIAN Bulgarica	CROATIAN Croatica	CZECH Bohemica	DANISH Danica	DUTCH Hollandica	FINNISH Fennica
brajlov šrift (брайлов шрифт)	Brailleovo pismo	slepecké písmo Braillovo	blindeskrift	blindenschrift braillenschrift	sokeainkirjoitus pistekirjoitus
svatbena pesen (сватбена песен)	svadbena pjesma	svatební píseň	bryllupsdigt	bruiloftsgedicht	häälaulu
bjuletin (бюлетин) izvestija (известия) *pl.*	bilten	věstník bulletin	beretning bulletin	bulletin	tiedotuslehti tiedonanto(lehti)
ot (от)	od	od	af	van	-n

kartička (картичка)
fiš (фиш)
kartonče (картонче)
kataložno
  kartonče
  (каталожно
  картонче)
	listić,
	kataloški listić
	knjižni listić
		lístek
		záznam
		lístek katalogový
			(katalog)kort
			(katalog)seddel
				cataloguskaart	kortti
					luettelokortti

fišov katalog (фишов каталог)	katalog na listići- ma	lístkový katalog	seddelkatalog	cartotheek kaartcatalogus	kortisto
katalog (каталог)	katalog	katalog	katalog	catalogus	lista luettelo
katalogiziram (каталогизирам)	katalogizirati	katalogizovat	katalogisere	catalogiseren	luetteloida
katalogizacija (каталогизация) opisanie na knigi (описание на книги)	katalogizacija	jmenná katalogi- zace	katalogisering	catalogisering titelbeschrijving	luettelointi

		GREEK Graeca	HUNGARIAN Hungarica	ITALIAN Italica	LATIN Latina	NORWEGIAN Norvegica
52		graphe Braille (γραφή Braille)	vakírás Braille-írás	scrittura Braille stampa sistema Braille	scriptura Braille (pro caecis)	blindeskrift
53		epithalamion (ἐπιθαλάμιον) gamelia (γαμελία)	nászdal	epitalamio	carmen nuptiale epithalamium	bryllupsdikt
54		deltion (δελτίον)	bulletin, értesítő, közlöny, jelentés	bollettino	acta	bulletin beretning
55		hypo (ὑπό)	-tól,-től	da, di per	a, ab	av
56		deltion (δελτίον)	cédula, kártya katalóguscédula katalóguskártya	scheda	charta scida charta/scida catalogi	katalogkort katalogseddel
57		deltiokatalogos (δελτιοκατάλογος)	cédulakatalógus	catalogo a schede schedario	chartoteca	seddelkatalog
58		katalogos (κατάλογος)	katalógus	catalogo	catalogus	katalog
59		syntasso katalogon (συντάσσω κατάλογον) katagrapho (καταγράφω)	katalogizál	catalogare	in indicem referre catalogisare	katalogisere
60		katalogographesis (καταλογογράφησις)	címleírás	catalogazione	ars catalogisandi catalogisatio	katalogisering

POLISH Polonica	PORTUGUESE Portugallica	RUMANIAN Rumenica	SERBIAN Servica	SLOVAK Slovaca	SWEDISH Suecica
pismo dla niewidomych pismo brajlowskie	escrita braille	scriere braille/oar- bă alfabetul orbilor	Brajevo pismo (Брајево писмо)	slepecké písmo	blindskrift
epitalamium pieśń weselna	epitalámio	epitalam	svadbena pesma (свадбена песма)	svadobná pieseň	bröllopskväde
biuletyn	boletim	buletin	bilten (билтен)	vestník bulletin	rapport meddelande bulletin
od	de por	de	od (од)	od	av
karta	ficha	fişă de catalog fişă	kataloški listić (каталошки листић)	lístok lístok katalógový	kort
katalog kartkowy	catálogo em ficha	catalog de fişă	katalog na listićima (каталог на листићима)	lístkový katalóg	kortkatalog
katalog	catálogo	catalog	katalog (каталог)	katalóg	bokförteckning katalog bulletin
katalogować	catalogar	cataloga	katalogizovati (каталогизовати)	katalogizovať	katalogisera
katalogowanie	catalogação	catalogare catalogizaţie	katalogizacija (каталогизација)	katalogizačný záznam	katalogisering

	ENGLISH Anglica	FRENCH Gallica	GERMAN Germanica	RUSSIAN Russica	SPANISH Hispanica
61	ceased publication (publication discontinued)	publication qui a cessé de paraître	erscheint nicht mehr eingegangen	prekrativšeesja izdanie (прекратившееся издание) izdanie prekratilos' (издание прекратилось)	no se publicó más
62	century	siècle	Jahrhundert	stoletie (столетие) vek (век)	siglo
	chairman →president				
63	change[1] (alter)	changer modifier	ändern verändern	izmenjať (изменять) vidoizmenjať (видоизменять)	cambiar mudar alterar
64	change[2] (be changed/altered)	changer	sich (ver)ändern	izmenjaťsja (изменяться)	cambiar mudar
65	chapter	chapitre	Kapitel	glava (глава)	capítulo
66	chief (main principal)	chef général premier principal	Haupt- Ober-, ober-	glavnyj (главный) osnovnoj (основной)	general principal en jefe
67	children's book (juvenile book)	livre pour enfants	Kinderbuch	detskaja kniga (детская книга) kniga dlja detej (книга для детей)	libro infantil
	choose →select				
	chrestomathy →selection				
	circulation →lending				
68	city library (borough/municipal library)	bibliothèque municipale	Stadtbücherei Stadtbibliothek städtische Bibliothek	gorodskaja biblioteka (городская библиотека)	biblioteca municipal

BULGARIAN Bulgarica	CROATIAN Croatica	CZECH Bohemica	DANISH Danica	DUTCH Hollandica	FINNISH Fennica
spreli izdanija (спрели издания)	(publikacija) dalje ne izlazi publikacija koja je prestala izlaziti	vydání zastaveno	ophørt at udkomme gået ind	gestaakte uitgave	(ilmestymästä) lakannut julkaisu
stoletie (столетие) vek (век)	stoljeće vijek	století	århundrede	eeuw	vuosisata
smenjam (сменям) promenjam (променям)	mijenjati	měnit pozměnit přizpůsobit	forandre omdanne	veranderen wijzigen	muuntaa
menja se (меня се) promenjam se (променям се)	mijenjati se	změnit se	forandre sig	veranderen wijzigen zich	muuntua muuttua
glava (глава)	poglavlje zaglavlje	kapitola hlava	afsnit kapitel	hoofdstuk kapittel	kappale
glaven (главен)	glavni	hlavní vedoucí	chef hoved	hoofd	pää- yli-
detska kniga (детска книга)	knjiga za decu	dětská kniha	børnebog	kinderboek	lastenkirja
gradska biblioteka (градска библиотека)	gradska knjižnica	městská knihovna	stadsbibliotek	gemeentebiblio- theek stadsbibliotheek	kaupungin- kirjasto

	GREEK Graeca	HUNGARIAN Hungarica	ITALIAN Italica	LATIN Latina	NORWEGIAN Norvegica

61	diekopse ekdosin
	(διέκοψε ἔκδοσιν)
		megszűnt
		félbemaradt
			pubblicazione
			  cessata
			spento (relativo
			  a periodici)
				publicatio cessata	opphørt å utkomme
					gått inn

62	hekatontaeteris (ἑκατονταετηρίς)	évszázad század	secolo	saeculum	århundre
63	metaballo (μεταβάλλω) metapoieo (μεταποιέω)	változtat	cambiare mutare variare modificare	mutare variare	forandre modifisere omdanne omforme
64	metaballomai (μεταβάλλομαι)	változik	cambiarsi mutarsi modificarsi	mutari variari modificari	forandre seg
65	kephalaion (κεφάλαιον)	fejezet	capitolo	caput capitulum	avsnitt kapittel
66	archi- (ἀρχι-) kyrios (κύριος)	fő-	primo primario capo-	primus princeps	hoved sjef
67	paidikon biblion (παιδικόν βιβλίον)	gyermekkönyv	libro per bambini	liber puerilis	barnebok
68	demotike biblitheke (δημοτική βιβλιοθήκη)	városi könyvtár	biblioteca comunale	bibliotheca municipalis	bybibliotek

POLISH Polonica	PORTUGUESE Portugallica	RUMANIAN Rumenica	SERBIAN Servica	SLOVAK Slovaca	SWEDISH Suecica
wydawnictwo zawieszone	publicação finda	încetat	prestalo da izlazi (престало да излази)	už nevyjde vydávanie sa skončilo	upphört att utkomma
stulecie wiek	século	secol veac	stoleće (столеђе) vek (век)	storočie	århundrade sekel
zmieniać	alterar modificar mudar variar	modifica schimba	menjati (мењати)	meniť premeniť zmeniť	förändra modifiera
zmieniać się	trocar	schimba, se modifica, se	menjati se (мењати се)	(z)meniť sa	ändra sig förändra sig
rozdział	capítulo	capitol cap	poglavlje (поглавље) zaglavlje (заглавље)	kapitola	avdelning avsnitt kapitel
główny	principal superior	principal general -şef	glavni (главни)	hlavný	över- huvud-
książka dla dzieci	livro para crianças	carte pentru copii	knjiga za decu (књига за децу)	detská kniha	barnbok
biblioteka miejska	biblioteca municipal	bibliotecă municipală	gradska biblioteka (градска библиотека)	mestská knižnica	stadsbibliotek

	ENGLISH Anglica	FRENCH Gallica	GERMAN Germanica	RUSSIAN Russica	SPANISH Hispanica
69	**classification** (**classifying**)	classification	Klassifikation	klassifikacija (классификация)	clasificación
70	**clause**	note finale clause	Klausel	zaključenie (заключение) klauzula (клаузула)	cláusula
71	**codex**	codex	Kodex	kodeks (кодекс)	códice
72	**collaborate** (**cooperate assist**)	collaborer coopérer concourir	mitarbeiten mitwirken zusammenarbeiten	sotrudničať (сотрудничать)	colaborar cooperar
73	**collaborator**	collaborateur	Mitarbeiter	sotrudnik (сотрудник)	colaborador cooperador
74	**collect** (**gather**)	(r)assembler collecter recueillir	sammeln	sobirať (собирать)	colegir recoger compilar
	collected works →**complete works**				
75	**collection**	collection recueil	Sammlung	sbornik (сборник) sobranie (собрание)	colección
76	**collotype** (**phototype**)	phototypie	Lichtdruck	fototipija (фототипия)	fototipia
77	**column**	colonne	Kolumne Spalte	kolonka (колонка) stolbec (столбец) grafa (графа)	columna
78	**comedy**	comédie	Komödie Lustspiel	komedija (комедия)	comedia
	come out →*be* **published**				
79	**comic** (**news**)**paper** (**humorous newspaper**)	journal humoristique	Witzblatt	jumorističeskaja gazeta (юмористическая газета)	tebeo

BULGARIAN Bulgarica	CROATIAN Croatica	CZECH Bohemica	DANISH Danica	DUTCH Hollandica	FINNISH Fennica
klasifikacija (класификация)	klasifikacija stručni raspored	klasifikace	systematisk ordning	classificering rubricering	luokitus
klauza (клауза)	klauzula	doložka klauzula	klausul	clausule	klausuuli lisäpykälä
kodeks (кодекс)	kodeks	kodex	kodeks	codex	käsikirjoitus
sâtrudniča (сътруднича)	surađivati	spolupracovat spolupůsobit	medvirke	medewerken samenwerken	avustaa yhdessä
sâtrudnik (сътрудник)	suradnik	spolupracovník	medarbejder	medewerker	avustaja
sâbiram (събирам)	skupljati sabirati	sebrat shromáždit	samle	verzamelen	kerätä koota
sbornik (сборник) sbirka (сбирка)	zbirka zbornik	sborník sbírka	indsamling samling	collectie verzameling	kokoelma
fototipija (фототипия)	fototipija svjetlotisak	světlotisk	lystryk	lichtdruk	valopainate
kolona (колона)	stupac	sloupec	spalte	kolom	palsta
komedija (комедия)	komedija	komedie veselohra	komedie lystspil	blijspel	huvinäytelmä komedia
vestnik za humor i satira (вестник за хумор и сатира) humorističen vestnik (хумористичен вестник)	šaljivi list humoristični list	humoristický časopis	vittighedsblad	humoristisch blad	pilalehti

43

	GREEK Graeca	HUNGARIAN Hungarica	ITALIAN Italica	LATIN Latina	NORWEGIAN Norvegica
69	systematike taxino- mesis (συστηματική ταξινόμησις)	osztályozás	classificazione	classificatio	klassifisering systematisering
70	retra (ρήτρα)	záradék	clausola	clausula conclusio	klausul
71	kodix (κώδιξ)	kódex	codice	codex	kodeks
72	synergazomai (συνεργάζομαι)	közreműködik	collaborare cooperare	collaborare assistere cooperari	medarbeide medvirke
73	synergates (συνεργάτης)	munkatárs	collaboratore	adiuvans assistens collaborator	medarbeider
74	syllego (συλλέγω)	gyűjt összegyűjt	raccogliere	colligere contribuere	samle in samle sammen
75	sylloge (συλλογή) analekta (ἀνάλεκτα)	gyűjtemény	collezione raccolta	collectio thesaurus	samling
76	phototypia (φωτοτυπία)	fénynyomat	fototipia	phototypia	lystrykk
77	stele (στήλη)	hasáb	colonna	columna	spalte
78	komodia (κωμῳδία)	vígjáték	commedia	comœdia	komedie lystspill
79	satirikon periodikon (σατιρικόν περιοδικόν)	vicclap humorisztikus lap	giornale umoristico	periodica iocosa	vittighetsblad

44

POLISH Polonica	PORTUGUESE Portugallica	RUMANIAN Rumenica	SERBIAN Servica	SLOVAK Slovaca	SWEDISH Suecica
układ syste- matyczny	classificação	clasificaţie	klasifikacija (класификација)	systematické radenie	klassifikation
klauzula zastrzeżenie	cláusula	clauză	završetak (завршетак) klauzula (клаузула)	doložka klauzula	klausul
kodeks	códice	codice cod(ex)	kodeks (кодекс)	kódex	kodex
współpracować	colaborar	colabora	sarađivati (сарађивати)	spolupracovať	medarbeta medverka samarbeta
współpracownik	colaborador	colaborator	saradnik (сарадник)	spolupracovník	medarbetare
zbierać	coleccionar coligir	culege	skupljati (скупљати) sabirati (сабирати)	pozbierať zhromaždiť	samla
zbiór kolekcja	colecção	colecţie	zbirka (збирка) zbornik (зборник)	zbierka	samling
fototypia	fototipia	fototipie	fototipija (фототипија)	svetlotlač	ljustryck
szpalta	coluna	coloană	kolona (колона)	stĺpec	spalt
komedia	comédia	comedie	komedija (комедија)	komědia česelohra	komedi lustspel
czasopismo humorystyczne	jornal humorístico	revistă umoristică	humorističke novine (хумористичке новине) pl.	humoristický časopis	serietidning

		ENGLISH Anglica	FRENCH Gallica	GERMAN Germanica	RUSSIAN Russica	SPANISH Hispanica
80		comic strip (strip cartoon)	bande dessinée	Comic strip Comics	komiks (комикс)	historieta
81		comment[1] *v.*	commenter	erläutern	kommentirovať (комментировать) tolkovať (толковать)	comentar
		comment[2] *n.* →commentary, note[1]				
82		commentary (comment[2])	commentaire	Kommentar Erläuterung	kommentarij (комментарий) tolkovanie (толкование)	comentario
		commission →committee				
83		committee (commission)	comité commission	Komitee Kommission Ausschuß	komitet (комитет) komissija (комиссия)	comité comisión
84		compile	composer compiler	zusammenstellen	sostavljať (составлять) kompilirovať (компилировать)	compilar recopilar
85		complete *a.* (full, entire, whole)	complet entier	sämtlich voll(ständig)	komplektnyj (комлектный) polnyj (полный)	completo entero
86		complete *v.* (supplement)	compléter suppléer	ergänzen	dopolnjať (дополнять) dobavljať (добавлять)	completar
87		complete works *pl.* (collected works *pl.*)	œuvres complètes *pl.*	sämtliche Werke *pl.* gesammelte Werke *pl.* Gesamtausgabe	polnoe sobranie sočinenij (полное собрание сочинений)	obras completas *pl.*
		conclusion →epilogue				

BULGARIAN Bulgarica	CROATIAN Croatica	CZECH Bohemica	DANISH Danica	DUTCH Hollandica	FINNISH Fennica
komiks (комикс)	šaljivi/humoris-tični strip	comics	tegneserie	illustratie serie beeldverhaal	sarjakuva
komentiram (коментирам) razjasnjavam (разяснявам) tâlkuvam (тълкувам)	komentirati	komentovat	kommentere	ophelderen uitleggen	selittää
komentar (коментар) tâlkuvane (тълкуване)	objašnjenje komentar	komentář výklad	bemærkning kommentar	commentaar	selitys
komitet (комитет) komisija (комисия)	komisija odbor	výbor	komité kommission	comité commissie	komitea toimikunta
sâstavjam (съставям)	sastaviti složiti	sestavit kompilovat	konstruere	samenstellen	laatia panna kokoon
komplekten (комплектен) pâlen (пълен) cjal (цял)	čitav potpun	celek celý úplný	fuldstændig komplet	compleet volledig	koko(nainen)
dopâlnjam (допълням) komplektuvam (комплектувам)	dopuniti popuniti	doplnit	fuldstændiggøre komplettere	aanvullen voltooien	täydentää
pâlno izdanie (пълно издание) pâlno sâbranie na sâčinenijata (пълно събрание на съчиненията)	cjelokupna djela *pl.*	sebrané spisy *pl.* sebraná díla *pl.*	samlede værker *pl.*	volledige werken *pl.*	kootut teokset

	GREEK Graeca	HUNGARIAN Hungarica	ITALIAN Italica	LATIN Latina	NORWEGIAN Norvegica
80	eikonographemenon mythistorema (εἰκονογραφημένον μυθιστόρημα)	képszalagos folytatá- sos kalandos regény „strip"	„comic strip" striscia di fumetti	series illustrationum delectantium in diurnis	tegneserie
81	scholiazo (σχολιάζω) hermeneuo (ἑρμηνεύω) exego (ἐξηγῶ)	kommentál magyaráz	commentare spiegare	commentare	kommentere
82	scholion (σχόλιον) hermeneia (ἑρμηνεία) exegesis (ἐξήγησις)	magyarázat	commentario illustrazione spiegazione	commentarium explicatio explanatio illustratio interpretatio	kommentar bemerkning
83	epitrope (ἐπιτροπή)	bizottság	comitato commissione	comitatus	komité kommisjon
84	syntasso (συντάσσω)	összeállít	comporre compilare	construere componere compilare	sammenstille kompilere
85	pleres (πλήρης) holos (ὅλος)	teljes összes minden	tutto intero completo	omnis cunctus totus	komplett
86	symplerono (συμπληρώνω)	kiegészít	completare	supplere complere	fullstendiggjøre komplettere supplere
87	pleres sylloge ton syngrammaton (πλήρης συλλογή τῶν συγγραμμάτων) hapanta (ἅπαντα) pl.	összes/összegyűjtött művek pl.	opere complete pl.	opera omnia pl.	samlede verker pl.

POLISH Polonica	PORTUGUESE Portugallica	RUMANIAN Rumenica	SERBIAN Servica	SLOVAK Slovaca	SWEDISH Suecica
historyjka obrazkowa komiks	banda desenhada	,,comic strip''	strip (стрип)	comics	tecknad film
komentować	comentar	comenta	komentarisati (коментарисати)	komentovať	kommentera
objaśnienie wyjaśnienie komentarz interpretacja	comentário explicação	interpretare comentariu lămurire	objašnjenje (објашњење) komentar (коментар)	komentár výklad	kommentar
komitet komisja	comissão	comitet comisie	komisija (комисија) odbor (одбор)	výbor	kommission kommitté
zebrać zestawiać kompilować	compilar	întocmi alcătui	sastaviti (саставити) složiti (сложити)	zostaviť zložiť kompilovať	sammanställa
kompletny wszystek	completo todo tudo	tot complet	čitav (читав) potpun (потпун)	celý všetok úplný	samtlig all komplett
dopełniać uzupełniać	completar	întregi completa	dopuniti (допунити) popuniti (попунити)	doplniť kompletovať	komplettera
wydanie kom- pletne	edição completa obras completas	opere complete *pl.* ediție completă	celokupna dela (целокупна дела) *pl.*	zobrané spisy/ diela *pl.*	fullständig upplaga

	ENGLISH Anglica	FRENCH Gallica	GERMAN Germanica	RUSSIAN Russica	SPANISH Hispanica
88	conference	conférence	Konferenz	konferencija (конференция) soveščanie (совещание)	conferencia
89	congress	congrès	Kongreß	s"jezd (съезд) kongress (конгресс)	congreso
90	contain	contenir comprendre	enthalten	soderžať (содержать)	contener comprender
91	contents pl. (argument)	contenu sommaire	Inhalt	soderžanie (содержание)	contenido
92	continue	continuer	fortsetzen	prodolžať (продолжать)	continuar
93	contribution (data pl.)	contribution	Beitrag	dannye (данные) pl.	contribución adición
	controversial pamphlet →polemic				
	cooperate →collaborate				
94	copperplate	chalcotypie impression en taille-douce	Kupferdruck	ofort (офорт) glubokaja pečať s mednych form (глубокая печать с медных форм)	calcotipia
95	copperplate engraving	taille-douce gravure en taille- douce	Kupferstich	gravjura na medi (гравюра на меди) ėstamp (эстамп)	grabado en cobre estampa
	copperplate printing →copperplate				
96	copy[1] (transcript)	copie	Kopie Abschrift	kopija (копия)	copia duplicado
97	copy[2] (exemplar, number[2])	exemplaire	Exemplar	ėkzempljar (экземпляр)	ejemplar

BULGARIAN Bulgarica	CROATIAN Croatica	CZECH Bohemica	DANISH Danica	DUTCH Hollandica	FINNISH Fennica
konferencija (конференция) sâveštanie (съвещание)	konferencija	konference	konference	conferentie	kokous
kongres (конгрес)	kongres	sjezd kongres	kongres	congres	kongressi
sâdâržam (съдържам)	sadržavati	obsahovat	indeholde	bevatten inhouden	sisältää
sâdâržanie (съдържание)	sadržaj	obsah	indhold	inhoud	sisällys sisältö
prodâlžavam (продължавам)	nastavljati	pokračovat	forsætte	voortzetten	jatkaa
prinos (принос)	prinos podaci *pl.*	příspěvek	bidrag	bijdrage	liite lisä(tieto)
halkografija (халкография) medna gravjura (медна гравюра)	bakrotisak	lept	kobbertryk	kopergravure	kuparipaino
rezba vârhu med (резба върху мед)	bakrorez	mědirytina	kobberstik	kopergravure	kuparipiirros vaskipiirros
prepis (препис)	kopija	kopie opis	kopi	afschrift kopie	kopio jäljennös
ekzempljar (екземпляр) brojka (бройка)	primjerak egzemplar	exemplář výtisk	eksemplar	exemplaar	kappale eksemplaari

		GREEK Graeca	HUNGARIAN Hungarica	ITALIAN Italica	LATIN Latina	NORWEGIAN Norvegica
88		syndiaskepsis (συνδιάσκεψις)	értekezlet konferencia	conferenza	conferentia	konferanse
89		synedrion (συνέδριον) kongresson (κογγρέσσον)	kongresszus	congresso	congressus	kongress
90		periecho (περιέχω)	tartalmaz	contenere	continere	inneholde
91		periechomenon (περιεχόμενον)	tartalom	contenuto	argumentum summa	innhold
92		synechizo (συνεχίζω)	folytat	continuare	continuare	fortsette
93		symbole (συμβολή)	adalék	contributo	contributum additamentum addenda	bidrag
94		chalkographema (χαλκογράφημα)	réznyomat réznyomtatás	stampa in rame	imago per aëneam laminam expressa	kobberstikk
95		chalkotechnia (χαλκοτεχνία) chalkographia (χαλκογραφία)	rézmetszet rézmetszés réznyomás	incisione in rame	chalcographia imago in aere incisa	kobberstikk
96		antigraphon (ἀντίγραφον)	másolat átírás	copia duplicato	apographum duplicatum	avskrift kopi
97		antitypon (ἀντίτυπον)	példány	esemplare copia	exemplar exemplum	eksemplar

POLISH Polonica	PORTUGUESE Portugallica	RUMANIAN Rumenica	SERBIAN Servica	SLOVAK Slovaca	SWEDISH Suecica
konferencja narada	conferência	consfătuire conferinţă	konferencija (конференција)	konferencia porada	konferens sammanträde
zjazd kongres	congresso	congres	kongres (конгрес)	zjazd kongres	kongress
obejmować zawierać	compreender conter	conţine cuprinde	sadržavati (садржавати)	obsahovať	innehålla
treść zawartość	contenudo	conţinut cuprins	sadržaj (садржај)	obsah	innehåll
kontynuować	continuar	continua	nastavljati (настављати)	pokračovať	fortsätta
przyczynek	contribuição artigo	contribuţie	prinos (принос) podaci (подаци) *pl.*	príspevok	bidrag
miedzioryt	gravura em cobre	gravură pe cupru calcotipie stampă	bakrotisak (бакротисак)	meditlač	koppartryck
miedzioryt rycina	gravura em talhe doce	gravură stampă	bakrorez (бакрорез)	medirytina	kopparstick
kopia odpis	cópia	copie duplicat	kopija (копија)	kópia odpis	avskrift kopia
egzemplarz	exemplar	exemplar	primerak (примерак) egzemplar (егземплар)	exemplár výtlačok	exemplar

	ENGLISH Anglica	FRENCH Gallica	GERMAN Germanica	RUSSIAN Russica	SPANISH Hispanica
98	**copyright**	droit d'auteur	Urheberrecht	avtorskoe pravo (авторское право)	derecho de autor
99	**copyright deposit**	dépôt légal	gesetzliche Pflichtablieferung Ablieferungspflicht	predostavlenie objazatelnogo ėkzempljara (предоставление обязательного экземпляра)	depósito legal
100	**corporate author**	collectivité-auteur collectivité d'auteurs	korporativer/ körperschaftlicher Verfasser Kollektivverfasser	kollektivnyj avtor (коллективный автор) kollektiv avtorov (коллектив авторов)	autor corporativo
101	**correct**	corriger améliorer	verbessern	ispravljať (исправлять) korrektirovať (корректировать)	corregir rectificar enmendar
102	**correction** **(proof)**	correction	Korrektur	korrektura (корректура)	corrección
103	**correspondence**	correspondance	Briefwechsel	korrespondencija (корреспонденция) perepiska (переписка)	correspondencia
	criticism →**critique**				
104	**critique** **(criticism)**	critique	Kritik	kritika (критика)	crítica
105	**criticize**	critiquer	besprechen rezensieren	kritikovať (критиковать)	criticar
	cross-reference card →**reference card** **cut** →**engraving** **cyclopaedia** →**encyclopaedia** **data** *pl.* →**contribution**				

BULGARIAN Bulgarica	CROATIAN Croatica	CZECH Bohemica	DANISH Danica	DUTCH Hollandica	FINNISH Fennica
avtorsko pravo (авторско право)	autorsko pravo	autorské právo	ophavsret	auteursrecht	tekijänoikeus
zadâlžitelno depozirane (задължително депозиране)	obveza dostave tiskanih stvari	povinná dodávka	pligtaflevering	wettelijk depot	lakimääräiset vapaakappale-luovutukset
kolektiven avtor (колективен автор) avtorski kolektiv (авторски колектив)	kolektivni autor	korporativní autor	korporativ forfatter	corporatieve auteur	ryhmätekijä
popravjam (поправям)	popraviti ispraviti	opravit opravovat	forbedre	verbeteren	korjata oikaista
popravka (поправка) korekcija (корекция)	korektura ispravka	opravy korektura	korrektur	correctie verbetering	korjaus oikaisu korrehtuuri
korespondencija (кореспонденция) prepiska (преписка)	dopisivanje	dopisování korespondence	brevveksling korrespondance	briefwisseling correspondentie	kirjeenvaihto
kritika (критика) otziv (отзив)	kritika	kritika	kritik	kritiek	arvostelu
kritikuvam (критикувам) razglеždam (разглеждам)	ocjenjivati kritizirati	kritizovat posuzovat	kritisere	critiseren	arvostella

	GREEK Graeca	**HUNGARIAN** Hungarica	**ITALIAN** Italica	**LATIN** Latina	**NORWEGIAN** Norvegica
98	syngraphikon dikaioma *(συγγραφικόν δικαίωμα)*	szerzői jog	diritto d'autore	ius auctoris	forfatterrett
99	kata nomon prosphora *(κατά νόμον προσφορά)*	kötelespéldány beszolgáltatása	deposito legale	officium tradendi exemplaris obligati	avleveringsplikt
100	syllogike ekdosis *(συλλογική ἔκδοσις)*	testületi szerző szerzői munka-közösség	opera di ente collettivo ente collettivo	auctor corporativus	korporativ forfatter
101	diorthono *(διορθώνω)*	javít	correggere emendare	corrigere emendare	forbedre
102	diorthose *(διόρθωση)*	javítás	correzione	correctio	korrektur
103	allelographia *(ἀλληλογραφία)* grammata *(γράμματα) pl.*	levelezés levélváltás	corrispondenza carteggio	correspondentia epistularum commercium	brevveksling korrespondanse
104	kritike *(κριτική)*	bírálat kritika	critica recensione	recensio	kritikk
105	epikrino *(ἐπικρίνω)*	bírál	criticare	recensere	kritisere

POLISH Polonica	PORTUGUESE Portugallica	RUMANIAN Rumenica	SERBIAN Servica	SLOVAK Slovaca	SWEDISH Suecica
prawo autorskie	direitos de autor propriedade intelectual	drept de autor	autorsko pravo (ауторско право)	autorské právo	författarrätt
prawo egzem- plarza obowiąz- kowego	depósito legal	obligaţiune de predare	predaja obaveznog primerka (предаја обавезног примерка)	povinnosť odovzdania povinného výtlačku	leverans av pliktexemplar
autor korpora- tywny	colectividade- -autor	autor corporativ	kolektivni autor (колективни аутор)	korporatívny autor	korporativ förfat- tare
poprawiać	corrigir emendar	corecta îndrepta	ispraviti (исправити) popraviti (поправити)	opraviť	förbättra
poprawka korekta	correcção emenda	corectură îndreptare	korektura (коректура) ispravka (исправка)	oprava korektúra	korrigering korrektur
korespondencja	correspondência	corespondenţă	korespondencija (кореспонденција) prepiska (преписка)	korešpondencia	brevväxling korrespondens
krytyka recenzja	crítica	critică	kritika (критика)	kritika	kritik
krytykować recenzować	criticar	critica recenza	ocenjivati (оцењивати) kritikovati (критиковати)	posudzovať recenzovať	bedöma kritisera

		ANGLICA Anglica	FRENCH Gallica	GERMAN Germanica	RUSSIAN Russica	SPANISH Hispanica
106		date of printing	date d'impression	Druckjahr	god pečatanija (год печатания)	fecha de impresión
107		decimal classification	classification décimale	Dezimalklassifikation	desjatičnaja klassifikacija (десятичная классификация)	clasificación decimal
		decree →order				
108		dedicated	dédié dédicacé	gewidmet	posvjaščĕnnyj (посвящённый)	dedicado
109		dedication (inscription)	dédicace envoi d'auteur	Widmung	posvjaščenie (посвящение)	dedicación dedicatoria
110		defective	défectueux	defekt	defektnyj (дефектный)	defectuoso
111		defend	défendre	verteidigen	zaščiščať (защищать)	defender
112		deposit copy	exemplaire du dépôt légal	Freiexemplar Pflichtexemplar	objazateĺnyj ékzempljar (обязательный экземпляр)	ejemplar de depósito legaı
		description →catalog(u)ing				
113		desiderata *pl*	desiderata *pl.*	Desiderat(um)	dezideraty (дезидераты) *pl.*	desiderata *pl.*
114		dialog(ue)	dialogue	Dialog	dialog (диалог)	diálogo
115		diary (journal)	journal	Tagebuch	dnevnik (дневник) žurnal (журнал)	diario
116		dictionary (lexicon)	dictionnaire vocabulaire	Wörterbuch	slovar' (словарь)	diccionario vocabulario

BULGARIAN Bulgarica	CROATIAN Croatica	CZECH Bohemica	DANISH Danica	DUTCH Hollandica	FINNISH Fennica
data na otpečatvaneto (дата на отпечатването)	godina tiska	vročení	trykår	drukjaar	painovuosi
desetična klasifikacija (десетична класификация)	decimalna klasifi- kacija	desetinné třídění	decimalklassi- fikation	decimale classificatie	kymmenluokitus
posveten (посветен)	posvećen	věnovaný	dediceret	opgedragen	omistettu
posveštenie (посвещение)	posveta	věnování	dedikation tilegnelse	opdracht	omistus omistuskirjoitus
defekten (дефектен)	defektan	defektní vadný	ufuldstændig ukomplet	incompleet onvolledig	epätäydellinen puutteellinen
zaštištavam (защищавам)	štititi zaštititi	obhájit	forsvare	verdedigen	puolustaa
zadâlžitelen depoziten ekzempljar (задължителен депозитен екземпляр)	obvezan primje- rak	povinný výtisk	pligtaflevrings- eksemplar	wettelijk verplicht exemplaar presentexemplaar	vapaakappale
deziderata (дезидерата)	deziderat	desiderata	desiderata	desiderata wensen	puuttuva teos
dialog (диалог)	dijalog razgovor	dialog rozmluva rozprava	dialog	dialoog samenspraak	dialogi vuoropuhelu
dnevnik (дневник)	dnevnik	deník	dagbog	dagboek	päiväkirja
rečnik (речник)	rječnik	slovník	ordbog	woordenboek	sanakirja

	GREEK Graeca	HUNGARIAN Hungarica	ITALIAN Italica	LATIN Latina	NORWEGIAN Norvegica
106	chronologia ektyposeos (χρονολογία ἐκτυπώσεως)	a kiadvány keltezése/ kelte kiadás éve	anno di stampa	annus imprimendi/ /impressionis	trykkår
107	dekadike taxinomesis (δεκαδική ταξινόμησις)	tizedes osztályozás	classificazione decimale	classificatio decimalis	desimalklassifikasjon
108	aphieromenos (ἀφιερωμένος)	ajánlással dedikálva	dedicato	dedicatus	dedisert
109	aphieroma (ἀφιέρωμα) aphierosis (ἀφιέρωσις)	ajánlás	dedica dedicazione	commendatio dedicatio	dedikasjon tilegnelse
110	elattomatikos (ἐλαττωματικός)	hibás	difettoso	defectus	defekt
111	hyperaspizo (ὑπερασπίζω)	megvéd	difendere	defendere	forsvare
112	antitypon tes kata nomon prosphoras (ἀντίτυπον τῆς κατά νόμον προσφορᾶς)	kötelespéldány	esemplare d'obbligo	exemplar obligatum	pliktavleverings- eksemplar
113	desiderata (δεσιδεράτα)	desideratum	desiderata	desideratum	desiderata *pl.*
114	dialogos (διάλογος)	párbeszéd	dialogo	colloquium dialogus sermo	dialog samtale
115	hemerologion (ἡμερολόγιον)	napló	diario giornale	commentarii diurni diarium	dagbok
116	lexikon (λεξικόν)	szótár	dizionario vocabolario	lexicon dictionarium thesaurus verborum vocabularium	ordbok

POLISH Polonica	PORTUGUESE Portugallica	RUMANIAN Rumenica	SERBIAN Servica	SLOVAK Slovaca	SWEDISH Suecica
data druku	data de impressão	dată de tipar	godina štampanja (година штампања)	rok tlače	tryckår
klasyfikacja dziesiętna	classificação decimal	clasificaţie decimală	decimalna klasifikacija (децимална класификација)	desatinné triedenie	decimalklassifika-tion
dedykowany	dedicado	dedicat	posvećen (посвећен)	venovaný	dedicerad
poświęcenie dedykacja	dedicatória	dedicaţie	posveta (посвета)	venovanie	dedikation
zdefektowany	defeituoso	defect defectat	defektan (дефектан)	defektný chybný	defekt
bronić	defender	apăra	štititi (штитити) zaštititi (заштитити)	obhájiť	försvara
egzemplarz obowiązkowy	exemplar de depó-sito legal	exemplar de obli-gaţie	obavezan primerak (обавезан примерак)	povinný výtlačok	pliktexemplar arkivexemplar
dezyderat	desiderata *pl.*	deziderat	deziderat (дезидерат)	deziderátum	desiderata
dialog	diálogo	dialog	dijalog (дијалог) razgovor (разговор)	dialóg	dialog samtal
diariusz dziennik	jornal	jurnal	dnevnik (дневник)	denník	dagbok
słownik dykcjonarz	dicionário vocabulário	dicţionar vocabular	rečnik (речник)	slovník	ordbok lexikon

		ENGLISH Anglica	FRENCH Gallica	GERMAN Germanica	RUSSIAN Russica	SPANISH Hispanica
		discuss →treat				
		dissertation →thesis				
117		document (record[2] file)	acte	Akt(e) Aktenstück Urkunde	dokument (документ) akt (акт)	documento acta
118		documentation	documentation	Dokumentation	dokumentacija (документация)	documentación
119		doubtful (dubious)	douteux	zweifelhaft	somnitel'nyj (сомнительный) nedostovernyj (недостоверный)	dudoso
		drama →play				
120		drawing	dessin	Zeichnung	risunok (рисунок)	dibujo diseño
		dubious →doubtful				
121		duplicate	double duplicata	Doppelstück Dublette Zweitexemplar	dublet (дублет) dublikat (дубликат)	duplicado
		duplication →reproduction				
122		edit[1] (prepare)	editer rediger	bearbeiten besorgen abfassen die Ausgabe besorgen	podgotavlivat' k pečati (подготавливать к печати)	editar preparar
123		edit[2] (publish)	éditer publier faire paraître	herausgeben	izdavat' (издавать) opublikovyvat' (опубликовывать)	editar publicar
124		edition (issue publication)	édition publication	Ausgabe Auflage	izdanie (издание) publikacija (публикация)	edición publicación
125		editor[1]	rédacteur	Redakteur	redaktor (редактор)	redactor

BULGARIAN Bulgarica	CROATIAN Croatica	CZECH Bohemica	DANISH Danica	DUTCH Hollandica	FINNISH Fennica
delo (дело) akt (акт) dokument (документ)	akt dokument isprava	akt listina spis dokument	dokument	akte document notulen *pl.* papieren *pl.*	asiakirja
dokumentacija (документация)	dokumentacija	dokumentace	dokumentation	documentatie	kirjallisuuspalvelu
sâmnitelen (съмнителен)	sumnjiv nevjerodostojan	nejistý pochybný	tvivlsom	twijfelachtig	epäil(lyt)tävä
risunka (рисунка) skica (скица)	crtež	kresba obrázek	tegning	tekening	piirustus
dublikat (дубликат)	duplikat prijepis	duplikát	duplikat	duplicaat kopie	kaksoiskappale duplikaatti

prigotvjam za
  pečat
(приготвям за
  печат)
redaktiram
(редактирам)
	prirediti	upravit
		připravit do
		  tisku
			drage omsorg for	zorgen voor . . .	järjestää
					panna kuntoon

izdavam (издавам)	izdati	vydat	udgive	uitgeven publiceren	julkaista
izdanie (издание)	izdanje izdavanje	náklad vydání	oplag udgave	editie uitgave druk	julkaisu laitos painos
redaktor (редактор)	urednik	redaktor	redaktør	bewerker redacteur	toimittaja

	GREEK Graeca	HUNGARIAN Hungarica	ITALIAN Italica	LATIN Latina	NORWEGIAN Norvegica
117	engrafa (ἔγγραφα) pl.	akta okmány	atto documento	acta pl. documentum scriptum	dokument
118	epicheirematologia (ἐπιχειρηματολογία) tekmeriosis (τεκμηρίωσις)	dokumentáció	documentazione	documentatio	dokumentasjon
119	amphibolos (ἀμφίβολος)	kétes	dubbio(so) incerto	dubius incertus	tvilsom
120	ichnographema (ἰχνογράφημα)	rajz	disegno	figura pictura linearis delineatio graphis	tegning
121	diplographon (διπλόγραφον)	másodpéldány duplum	duplicato	duplum	dublett duplikat gjenpart
122	epimeleomai (ἐπιμελέομαι)	gondoz sajtó alá rendez szerkeszt	curare	adcurare curare curam gerere	forberede til trykning
123	ekdido (ἐκδίδω)	kiad közzétesz	pubblicare	edere publicare in lucem ferre	forlegge utgive
124	ekdosis (ἔκδοσις)	kiadás	edizione pubblicazione	editio	publikasjon utgave
125	syntaktes (συντάκτης)	szerkesztő	redattore curatore	redactor	redaktør

64

POLISH Polonica	PORTUGUESE Portugallica	RUMANIAN Rumenica	SERBIAN Servica	SLOVAK Slovaca	SWEDISH Suecica
akt dokument	acta auto documento	act document	akt (акт) dokument (документ) isprava (исправа)	úradný spis akt dokument spis	dokument handling
dokumentacja	documentação	documentaţie	dokumentacija (документација)	dokumentácia	dokumentation
wątpliwy	dúbio duvidoso	dubios îndoielnic nesigur	sumnjiv (сумњив) neverodostojan (невородостојан)	neistý pochybný	tvivelaktig
rysunek szkic	debuxo desenho	desen	crtež (цртеж)	kresba nákres	teckning
dublet	duplicado	dublet	duplikat (дупликат) prepis (препис)	duplikát	dubblett
załatwić pielęgnować	preparar para a impressão	îngriji	prirediti (приредити)	upraviť pripraviť do tlače	besörja förbereda
wydawać	editar publicar	edita publica	izdati (издати)	vydať vydávať	utgiva publicera
edycja wydanie	edição	ediţie	izdanje (издање) izdavanje (издавање)	náklad vydanie	upplaga utgåva
redaktor	redactor	redactor	redaktor (редактор)	redaktor	redaktör

	ENGLISH Anglica	FRENCH Gallica	GERMAN Germanica	RUSSIAN Russica	SPANISH Hispanica
126	editor² (publisher)	editeur	Herausgeber	izdateľ (иядатель)	editor
127	editorial office (editor's office)	rédaction	Redaktion Schriftleitung	redakcija (редакция)	redacción
	editor's office →editorial office				
128	elaborate (work out/up)	élaborer traiter	ausarbeiten	razrabatyvaľ (разрабатывать)	elaborar
129	elegy	élégie	Elegie Klagelied	èlegija (элегия)	elegía
	employ →use v.				
130	encyclopaedia (cyclopaedia)	encyclopédie	Enzyklopädie Lexikon	ènciklopedija (энциклопедия)	enciclopedia
131	engrave	graver	gravieren	gravirovaľ (гравировать)	grabar
132	engraving (cut)	gravure estampe	Stich Bilddruck	gravjura (гравюра)	grabado
133	enlarge	augmenter	vermehren erweitern	rasširjaľ (расширять) uveličivaľ (увеличивать)	aumentar acrecentar ampliar
	entire→complete a. entry→catalog(u)ing				
134	epic (epos)	poéme épique épopée	Heldendichtung Heldengedicht Epos	èpičeskaja poema (эпическая поэма) èropeja (эпопея) èpos (эпос)	poema épico epopeya

BULGARICAN Bulgarica	CROATIAN Croatica	CZECH Bohemica	DANISH Danica	DUTCH Hollandica	FINNISH Fennica
izdatel (издател)	izdavač nakladnik	nakladatel vydavatel	udgiver	uitgever	julkaisija
redakcija (редакция)	uredništvo redakcija	redakce redaktorství	redaktion	redactie	toimitus
izrabotvam (изработвам) razrabotvam (разработвам)	razraditi obraditi	vypracovat zpracovat	udarbejde	bewerken uitwerken	muokata laatia
elegija (елегия)	elegija	elegie žalozpěv	elegie klagedigt	elegie klaagdicht treurzang treurdicht	elegia surulaulu valitusruno
enciklopedija (енциклопедия)	enciklopedija	encyklopedie naučný slovník	encyklopædi konversations- leksikon leksikon	encyclopedie lexicon	tietosanakirja ensyklopedia
graviram (гравирам)	gravirati rezati urezivati	rýt	indgravere	graveren	kaivertaa
gravjura (гравюра)	gravira rez	rytina	gravering stik	gravure prent	kaiverrus
uveličavam) (увеличавам) razširjavam (разширявам)	umnožiti proširiti	rozšířit zvětšovat	forstørre udvide	vermeerderen	laajentaa lisätä
epopeja (епопея) epična poema (епична поема)	epopeja ep	epos hrdinská báseň	epope heltedigt	epos heldendicht	eepos sankariruno

		GREEK Graeca	HUNGARIAN Hungarica	ITALIAN Italica	LATIN Latina	NORWEGIAN Norvegica
126		ekdotes (ἐκδότης)	kiadó	editore	editor	utgiver
127		syntaxis (σύνταξις)	szerkesztőség	redazione	redactio	redaksjon
128		exergazomai (ἐξεργάζομαι) syngrapho (συγγράφω)	feldolgoz	elaborare elucubrare	elaborare conficere elucubrare	bearbeide
129		elegeia (ἐλεγεία) epitymbidion (ἐπιτυμβίδιον)	elégia	elegia	cantus lugubris elegia nenia	elegi klagesang
130		enkyklopaideia (ἐγκυκλοπαίδεια) lexikon (λεξικόν)	lexikon enciklopédia	enciclopedia lessico	encyclopaedia lexicon	konversasjonslek- sikon
131		glypho (γλύφω)	metsz vés	incidere	sculpere incidere	gravere inngravere skjære
132		chalkographia (χαλκογραφία) xylographia (ξυλογραφία) glyphe (γλύφη)	metszet	incisione stampa	imago incisa sculptura	gravering kobberstikk
133		auxano (αὐξάνω) euryno (εὐρύνω)	bővít	amplificare	amplificare augere	utvide
134		epos (ἔπος) epopoiia (ἐποποιΐα)	hősköltemény eposz	epopea poema epico	epos epopoea	epos heltedikt

POLISH Polonica	PORTUGUESE Portugallica	RUMANIAN Rumenica	SERBIAN Servica	SLOVAK Slovaca	SWEDISH Suecica
wydawca	editor	editor	izdavač (издавач)	nakladateľ vydavateľ	utgivare
redakcja	redacção	redacţie	redakcija (редакција)	redakcia	redaktion
opracować	elaborar trabalhar	elabora prelucra	razraditi (разрадити) obraditi (обрадити)	vypracovať spracovať	utarbeta
elegia pieśń żałobna	elegia	elegie	elegija (елегија)	elégia žalospev	elegi
encyklopedia	enciclopédia	enciclopedie	enciklopedija (енциклопедија)	encyklopédia lexikón náučný slovník	encyklopedi konversations- lexikon
rytować	gravar	grava	gravirati (гравирати) rezati (резати) urezivati (урезивати)	rezať ryť	gravera
rycina sztych	gravura	gravură stampă	gravira (гравира) rez (рез)	rez rytina	gravyr stick
powiększyć rozszerzyć	alargar ampliar aumentar	adăuga mări lărgi	umnožiti (умножити) proširiti (проширити)	rozšíriť zväčšovať	utvidga
epopeja poemat bohaterski	epopeia	epopee poem epic	epopeja (епопеја) ep (еп)	epopeja hrdinská báseň hrdinský epos	epos hjältedikt

	ANGLICA Anglica	FRENCH Gallica	GERMAN Germanica	RUSSIAN Russica	SPANISH Hispanica
135	epilogue (conclusion)	postface épilogue	Nachwort Schlußwort Epilog	posleslovie (послесловие) zaključenie (заключение) ėpilog (эпилог)	epílogo
	epistle →letter				
	epos→epic				
136	essay[1]	essai	Aufsatz Essay	ėsse (эссе) očerk (очерк)	ensayo
	essay[2] →study, treatise				
	etched copperplate →etching				
137	etching (etched copperplate)	gravure à l'eau forte	Kupferradierung	gravjura na medi (гравюра на меди) ofort (офорт)	grabado al agua fuerte aguafuerte
	excerpt→extract				
138	exchange centre (exchange center)	service/centre des échanges	Tauschstelle	centr knigoobmena (центр книгообмена)	centro/servicio de canje
	exemplar→copy[2]				
139	exhibition (exposition)	exposition	Ausstellung	vystavka (выставка)	exhibición exposición
140	explain (interpret)	expliquer interpréter	erläutern	ob"jasnjať (объяснять) tolkovať (толковать)	explicar interpretar
141	explanation (explication)	interprétation explication	Erläuterung	ob"jasnenie (объяснение) tolkovanie (толкование)	explicación interpretación
	explanatory note →annotation				
	explication →explanation				
	exposition→exhibition				
142	extract (excerpt)	extrait abrégé	Auszug	izvlečenie (извлечение) vyderžka (выдержка)	extracto compendio epítome

70

BULGARIAN Bulgarica	CROATIAN Croatica	CZECH Bohemica	DANISH Danica	DUTCH Hollandica	FINNISH Fennica
posleslovie (послесловие) epilog (епилог)	pogovor epilog	doslov epilog	efterskrift epilog slutord	epiloog nawoord slotwoord	jälkilause epilogi loppulause
ese (есе)	esej ogled	essay	essay	essay	tutkielma kirjoitelma essee
gravjura (гравюра) halkografija (халкография)	bakropis	lept mědirytina	radering	ets	syövytys etsaus
knigoobmenna služba (книгообменна служба)	služba zamjene knjiga	výměnná služba	udvekslings- tjeneste	ruilbureau ruildienst	vaihtokeskus
izložba (изложба)	izložba	výstava	udstilling	tentoonstelling	näyttely
objasnjavam (обяснявам) razjasnjavam (разяснявам)	tumačiti objašnjavati	vykládat vyložit	forklare	uitleggen verklaren	selittää
objasnenie (обяснение)	objašnjenje	výklad	bemærkning	verklaring	selitys
izvadka (извадка) izvlečenie (извлечение) rezjume (резюме)	izvod odlomak izvadak	výpis výtah	uddrag	uittreksel	ote

	GREEK Graeca	HUNGARIAN Hungarica	ITALIAN Italica	LATIN Latina	NORWEGIAN Norvegica
135	epilogos *(ἐπίλογος)* teleutaia rhemata *(τελευταῖα ῥήματα)*	utószó	epilogo	epilogus peroratio	epilog etterskrift etterord slutningsbemerkning
136	dokimion *(δοκίμιον)*	esszé	saggio	studium elaboratum elucubratio	essay
137	chalkographia *(χαλκογραφία)*	rézkarc	incisione all'acquaforte acquaforte	figura in aere irrasa	kopperradering
138	hyperesia antallagon *(ὑπηρεσία ἀνταλ- λαγῶν)*	csereközpont	servizio degli scambi	institutum librorum permutandi	utvekslingstjeneste
139	ekthesis *(ἔκθεσις)*	kiállítás	esposizione	exhibitio expositio	utstilling
140	exego *(ἐξηγῶ)*	magyaráz	spiegare esporre illuminare	explicare exponere illuminare illustrare	forklare
141	exegesis *(ἐξήγησις)* scholion *(σχόλιον)*	magyarázat	spiegazione	explanatio illustratio interpretatio	bemerkning kommentar
142	apospasma *(ἀπόσπασμα)* epitome *(ἐπιτομή)*	kivonat	epitome estratto sunto	extractum epitome excerptum	utdrag

POLISH Polonica	PORTUGUESE Portugallica	RUMANIAN Rumenica	SERBIAN Servica	SLOVAK Slovaca	SWEDISH Suecica
posłowie epilog	pós-data	epilog postfață	pogovor (поговор) epilog (епилог)	doslov záver	epilog slutord
esej	ensaio	eseu	esej (есеj) ogled (оглед)	esej	essay, essä uppsats
rycina	gravura em água-forte	gravură cu acvaforte acvaforte	bakropis (бакропис)	medirytina	radering etsning
biuro wymiany	centro/serviço de permuta	serviciul de schimbare	služba razmene publikacija (служба размене публикација)	výmena	bytesavdelning
wystawa	exposição	expoziţie	izložba (изложба)	výstava	utställning
objaśniać wyjaśniać	explicar	explica lămuri interpreta	tumačiti (тумачити) objašnjavati (објашњавати)	vykladať vysvetľovať	förklara
interpretacja	explicação	explicaţie lămurire interpretare	objašnjenje (објашњење)	výklad	förklaring
streszczenie wyciąg	excerto extracto	extras	izvod (извод) odlomak (одломак) izvadak (извадак)	výpis výťah	utdrag

	ENGLISH Anglica	FRENCH Gallica	GERMAN Germanica	RUSSIAN Russica	SPANISH Hispanica
143	**fable** **(tale)**	fable conte (de fées)	Märchen Fabel	skazka (сказка) basnja (басня)	fábula cuento
144	**facsimile**	fac-similé	Faksimile	faksimile (факсимиле)	facsímile
145	**fashion magazine**	journal de modes	Modezeitschrift Modeblatt	žurnal mod (журнал мод)	revista de modas
	festschrift **→memorial volume**				
146	**fiction**	nouvelles et contes belles-lettres œuvres d'imagination	Romanliteratur erzählende Dichtung Belletristik	belletristika (беллетристика) povestvovatelnaja hudc žestvennaja literatura (повествовательная художественная литература)	novelas y cuentos
	figure **→picture, number**				
	file→document				
147	**first**	premier	erste	pervyj (первый)	primer(o) primera
	first edition **→original edition**				
148	**flysheet** **(leaflet** **broadside)**	feuille volante	Flugblatt	listovka (листовка)	folleto hoja volante
149	**footnote**	note infrapaginale note au bas de la page	Fußnote	podstročnoe primečanie (подстрочное примечание) snoska (сноска)	nota al pie
150	**for**	pour	für	dlja (для) radi (ради) za (за)	para por
	foreword **→preface**				

BULGARIAN Bulgarica	CROATIAN Croatica	CZECH Bohemica	DANISH Danica	DUTCH Hollandica	FINNISH Fennica
prikazka (приказка) basnja (басня)	bajka basna priča	bajka pohádka	eventyr fabel	fabel	satu tarina
faksimile (факсимиле)	faksimil	faksimile	faksimile	facsimile	faksimile jäljennöspainos
modno spisanie (модно списание) moden žurnal (моден журнал)	modni časopis	módní časopis	modeblad modejournal	modejournal modeblad	muotilehti
beletristika (белетристика) hudožestvena proza (художествена проза)	beletristika lijepa književnost	próza romány a novely	prosadigtning	romanliteratuur verhalend proza	suorasanainen kaunokirjalli- suus kertomakirjalli- suus (proosa)kirjallisuus
pârvi (първи)	prvi	první prvý	først(e)	eerste	ensi(mäinen)
poziv (позив) hvârčašt list (хвърчащ лист)	letak	leták list	flyveskrift flyveblad	vlugschrift	lentokirjanen lentolehti
beležka pod linija (бележка под линия)	bilješka na dnu stranice	poznámka pod čarou	note	voetnoot	alaviite
za (за)	za radi zbog	pro	for til	om voor	-lle takia varten vuoksi

	GREEK Graeca	**HUNGARIAN** Hungarica	**ITALIAN** Italica	**LATIN** Latina	**NORWEGIAN** Norvegica
143	paramythi *(παραμύθι)* mythos *(μῦθος)*	mese	fiaba storiella	fabula	eventyr fabel sagn
144	faksimile *(φαξσίμιλε)* apeikasma *(ἀπείκασμα)*	fakszimile	facsimile	facsimile	faksimile
145	periodikon modas *(περιοδικόν μόδας)*	divatlap	rivista di moda	periodica **vestium** elegantium	moteblad motejournal
146	logotechnike pezogra- phia *(λογοτεχνική πεζογραφία)*	széppróza	romanzi e novelle	prosa pulchra	diktning på prosa fiksjon oppdiktning
147	protos *(πρῶτος)*	első	primo	primus princeps	første
148	prokeryxe *(προκήρυξη)* feig-bolan *(φέϊγ-βολάν)*	röplap	foglio volante	libellus	flygeblad
149	parapompe *(παραπομπή)*	lábjegyzet	nota in calce	annotatio in ima pagina	fotnote
150	gia *(γιά)* pros *(πρός)* hyper *(ὑπέρ)* pro *(πρό)*	-ért számára -ra, -re	per	pro propter	for

POLISH Polonica	PORTUGUESE Portugallica	RUMANIAN Rumenica	SERBIAN Servica	SLOVAK Slovaca	SWEDISH Suecica
bajka baśń	conto fábula	basm poveste	bajka (бајка) basna (басна) priča (прича)	bájka rozpávka	fabel saga berättelse
facsimile	fac-símile	facsimil	faksimil (факсимил)	faksimile	faksimile
żurnal	jornal de modas	jurnal de mode	modni žurnal (модни журнал)	módny časopis módy *pl.*	modetidskrift
beletrystyka	romances e novelas	beletristică în proză	beletristika (белетристика)	próza romány a novely *pl.*	skönlitteratur
pierwszy	primeiro principal	întîi prim	prvi (први)	prvý	första, förste
druk ulotny ulotka	folheto papel volante	foaie volantă	letak (летак)	leták	flygskrift flygblad
przypis pod tekstem uwaga	nota no pé de página	notă în subsol	beleška na dnu stranice (белешка на дну странице)	poznámka pod čiarou	not anmärkning (nederst på sidan)
dla przez za	a para por	pentru	za (за) radi (ради) zbog (због)	pre	för

	ENGLISH Anglica	FRENCH Gallica	GERMAN Germanica	RUSSIAN Russica	SPANISH Hispanica
151	forgery	falsification	Fälschung	falsifikacija (фальсификация) poddelka (подделка)	falsificación
	founded on →based on				
152	fragment	fragment	Fragment Bruchstück	fragment (фрагмент) otryvok (отрывок)	fragmento trozo
153	from	d' de par	von aus	ot (от)	de desde por
	full →complete gather →collect				
154	glossary (vocabulary)	glossaire vocabulaire	Glossar Wörterverzeichnis	spisok slov (список слов) glossarij (глоссарий)	glosario vocabulario
	gramophon record →record[1] grand →large				
155	guide(-book)	guide	Reisebuch Reiseführer	putevoditeľ (путеводитель)	guía
156	handbook (manual)	manuel	Handbuch	spravočnik (справочник) rukovodstvo (руководство)	manual
157	history	histoire	Geschichte	istorija (история)	historia
	humorous newspaper →comic (news)paper				
	hymn →anthem				
	hymnal →songbook				
	hymn-book →songbook				
158	iconography	iconographie	Ikonographie	ikonografija (иконография)	iconografía
159	idyll	idylle	Idylle Hirtengedicht	idillija (идиллия)	idilio

BULGARICAN Bulgarica	CROATIAN Croatica	CZECH Bohemica	DANISH Danica	DUTCH Hollandica	FINNISH Fennica
falšifikat (фалшификат) podpravka (подправка)	falsifikat krivotvorina	falsifikát padělek	falskneri forfalskning	vervalsing namaak	väärennös falsifikaatti
fragment (фрагмент) otkâslek (откъслек)	fragment odlomak	ragment zlomek úryvek	brudstykke fragment	fragment	fragmentti katkelma
ot (от)	od	oď	af	van	-lta, -ltä -sta, -stä
spisâk na dumi (списък на думи)	glosar	glosář	fortegnelse glosar ordliste	lijst glossarium woordenlijst	sanaluettelo sanasto
pâtevoditel (пътеводител)	vodič	průvodce	fører vejledning	gids leidraad	opas
râkovodstvo (ръководство) narâčnik (наръчник)	priručnik	příručka rukověť	håndbog	handboek	käsikirja
istorija (история)	historija povijest	dějiny historie	historie	geschiedenis historie	kertomus historia
ikonografija (иконография)	ikonografija	ikonografie	ikonografi	iconografie	ikonografia
idilija (идилия)	idila	idyla	hyrdedigt idyl	herdersdicht idylle	idylli

	GREEK Graeca	HUNGARIAN Hungarica	ITALIAN Italica	LATIN Latina	NORWEGIAN Norvegica
151	notheusis (νόθευσις) kibdeleia (κιβδηλεία)	hamisítvány	falsificazione	falsum	forfalskning
152	apospasma (ἀπόσπασμα)	töredék	frammento	fragmentum	bruddstykke fragment
153	apo (ἀπό) ek (ἐκ)	-tól, -től -ból, -ből	da di	a, ab de e, ex	av fra
154	onomastikon (ὀνομαστικόν)	szójegyzék	lista glossario	glossarium vocabularium	glossar ordbok ordliste
155	hodegos (ὁδηγός)	útikönyv útikalauz	guida	ductor	guide fører veiledning
156	encheiridion (ἐγχειρίδιον)	kézikönyv	manuale	compendium liber manualis manuale enchiridion	håndbok
157	historia (ἱστορία)	történet történelem	storia	historia res gestae	historie
158	eikonographia (εἰκονογραφία)	ikonográfia	iconografia	iconographia	ikonografi
159	eidyllion (εἰδύλλιον)	idill pásztorköltemény	idillio	idyllium poëma bucolicum	hyrdedikt idyll

POLISH Polonica	PORTUGUESE Portugallica	RUMANIAN Rumenica	SERBIAN Servica	SLOVAK Slovaca	SWEDISH Suecica
falsyfikat podróbka	falsificação falso	falsificare falsificaţie	falsifikat (фалсификат) krivotvorina (кривотворина)	falzifikát	förfalskning
fragment urywek	fragmento	gragment fracţiune	fragment (фрагмент) odlomak (одломак)	fragment zlomok úryvok	fragment brottstycke
od z(e)	de por	de din	od (од) iz (из)	od z	av
glosariusz	glossário	glosar	glosar (глосар)	glosár	ordlista
przewodnik	guia	ghid călăuză	putovođa (путовођа)	sprievodca	vägledning guide
podręcznik	manual	compendiu manual	priručnik (приручник)	príručka rukoväť	handbok uppslagsbok
historia	história	istorie	istorija (историја) povest (повест)	dejiny *pl.* história	historia
ikonografia	iconografia	iconografie	ikonografija (иконографија)	ikonografia	ikonografi
idylla	idílio	idilă	idila (идила)	idyla	idyll

	ENGLISH Anglica	FRENCH Gallica	GERMAN Germanica	RUSSIAN Russica	SPANISH Hispanica
160	**illustrate** **illustration** →picture	illustrer	illustrieren	illjustrirovať (иллюстрировать)	ilustrar
161	**imitation**	imitation copie	Nachbildung Nachahmung Imitation	imitacija (имитация) podražanie (подражание)	imitación copia
162	**impression** **(printing** **print²)**	impression	Abdruck	otpečatok (отпечаток) ottisk (оттиск)	impresión tiraje tirada
163	**imprint** **(printer's address)**	adresse bibliographique	Druckvermerk Erscheinungsvermerk Impressum	vychodnye dannye (выходные данные)	notas tipográficas pie de imprenta
164	**in**	à dans en	in	v, vo (в, во) na (на)	a en
165	**incomplete**	incomplet	unvollständig	nekomplektnyj (некомплектный) nepolnyj (неполный)	incompleto
166	**incunabula** *pl.*	incunables *pl.*	Inkunabel Wiegendruck	inkunabula (инкунабула)	incunable
167	**index** **(register)**	index registre	Register	ukazateľ (указатель) indeks (индекс)	índice
168	**initial** **inscription**→dedication	initiale	Initiale Anfangsbuchstabe	inicial (инициал)	inicial
169	**insert** **(intercalate)**	insérer intercaler	einfügen	vstavljať (вставлять) vključať (включать)	insertar

BULGARIAN Bulgarica	CROATIAN Croatica	CZECH Bohemica	DANISH Danica	DUTCH Hollandica	FINNISH Fennica
iljustriram (илюстрирам)	ilustrirati	ilustrovat vyobrazovat	illustrere	illustreren	kuvittaa valaista
imitacija (имитация)	podražavanje imitacija	napodobení	efterligning imitation	imitatie	jäljittely mukailu
otpečatâk (отпечатък)	otisak štampa tisak	otisk tisk	tryk(sag) oplag	druk	vedos
bibliografski adres (библиографски адрес) izdatelski danni (издателски данни) *pl.*	impresum	tisk otisk	impressum	bibliografisch adres impressum	julkaisutiedot
v (в) vâv (във) na (на)	u na	v ve	i på	in	-ssa -ssä
nepâlen (непълен) nekomplekten (некомплектен)	manjkav nepotpun	neúplný	ufuldstændig ukomplet	onvolledig incompleet	epätäydellinen puutteellinen
inkunabul (инкунабул) pârvopečatna kniga (първопечатна книга)	inkunabula prvotisak	inkunabule prvotisk	inkunabel vuggetryk	incunabel wiegedruk	inkunaabeli varhaispainos
indeks (индекс) ukazatel (указател)	indeks registar	ukazatel ukazovatel	indeks	index	hakemisto
inicial (инициал)	inicijali *pl.*	iniciála	initial	beginletter initiaal	initiaalit *pl.* alkukirjain
vmestvam (вмествам)	umetati prilagođavati	vložit vpravit	indføje indsætte	invoegen	liittää lisätä sovittaa

		GREEK Graeca	HUNGARIAN Hungarica	ITALIAN Italica	LATIN Latina	NORWEGIAN Norvegica
160	eikonographo (εἰκονογραφῶ)	illusztrál	illustrare	illustrare	illustrere
161	mimesis (μίμησις)	utánzat	imitazione	imitatio	etterlikning imitasjon
162	anatyposis (ἀνατύπωσις)	nyomat	tiratura	impressio impressum typus	avtrykk opplag preg trykk
163	ektyposis (ἐκτύπωσις)	impresszum	nota tipografica soscrizione	impressum	impressum
164	en (ἐν) eis (εἰς), se (σε)	-ba, -be -ban, -ben	in	in	i på
165	ell(e)ipes (ἐλλειπής) asymplerotos (ἀσυμπλήρωτος)	hiányos	incompleto scompleto	incompletus	ufulstendig ukomplett
166	typoma archaion (τύπωμα ἀρχαῖον)	ősnyomtatvány inkunábulum	incunabulo	incunabulum	inkunabel paleotyp vuggetrykk
167	katalogos (κατάλογος) pinakas (πίνακας)	mutató index	indicatore indice	index	index viser
168	archikon gramma (ἀρχικόν γράμμα)	inicialé	iniziale	initiale	initiale
169	paremballo (παρεμβάλλω)	beilleszt	inserire intercalare	inserere intercalare	innføre innsette

POLISH Polonica	PORTUGUESE Portugallica	RUMANIAN Rumenica	SERBIAN Servica	SLOVAK Slovaca	SWEDISH Suecica
illustrować	ilustrar	ilustra	ilustrovati (илустровати)	ilustrovať	illustrera
naśladownictwo imitacja	imitação	imitaţie	podražavanje (подражавање) imitacija (имитација)	imitácia napodobnenina	imitation
druk	impressão tiragem	imprimare tipărire	otisak (отисак) štampa (штампа)	odtlačok tlač	tryck(sak)
adres wydawniczy metryka książki	pé de imprensa	adresă bibliogra- fică	impresum (импресум)	impressum	tryckuppgift
w(e)	em	în	u (у) na (на)	v vo	i på
niekompletny	defeituoso	incomplet	nepotpun (непотпун) nekompletan (некомплетан)	neúplný	ofullständig
inkunabuł	incunábulo	incunabul	inkunabule (инкунабуле) pl. prvotisak (првотисак)	inkunábula prvotlač	inkunabel
indeks	index	indice	indeks (индекс) registar (регистар)	index ukazovateľ	index register
inicjał	inicial	iniţială	inicijali (иницијали) pl.	iniciála	initial
adjustować	inserir	insera intercala	umetati (уметати) prilagođavati (прилагођавати)	vložiť zaradiť	infoga

	ENGLISH Anglica	FRENCH Gallica	GERMAN Germanica	RUSSIAN Russica	SPANISH Hispanica
170	institute	institut	Institut Anstalt	institut (институт)	instituto
	instructions *pl.* →prescription				
	intercalate →insert				
	interpret →explain				
171	introduce	introduire	einleiten einführen	vvodiť (вводить)	introducir
172	introduction	introduction	Einleitung Einführung	vvedenie (введение) vstuplenie (вступление)	introducción
173	inventory	inventaire	Bestand Inventur	inventar' (инвентарь)	inventario
	issue→edition				
174	joint author	co-auteur	Mitverfasser	soavtor (соавтор)	coautor
	journal →diary, periodical				
	juvenile book →children's book				
175	language	langage langue	Sprache	jazyk (язык)	lengua idioma
176	large (big grand)	grand	groß	boľšoj (большой) velikij (великий)	gran(de)
177	law (statute[2])	loi	Gesetz	zakon (закон)	ley estatuto
178	leaf	feuillet	Blatt	list (лист)	hoja
	leaflet →flysheet				

BULGARIAN Bulgarica	CROATIAN Croatica	CZECH Bohemica	DANISH Danica	DUTCH Hollandica	FINNISH Fennica
institut (институт)	zavod institut	ústav institut	anstalt institut	instituut	instituutti laitos opisto
vâveždam (въвеждам)	uvoditi	uvádět	indføre introducere præsentere	inleiden	johdatta
uvod (увод) vâvedenie (въведение)	uvod	předmluva úvod	indledning introduktion præsentation	inleiding	johdanto
inventar (инвентар)	inventar	inventář	bestand	inventaris	ınventaari(o)
sâavtor (съавтор)	suautor	spoluautor	medforfatter	medewerker	tekijäkumppani
ezik (език)	jezik	jazyk	sprog	spraak taal	kieli
goljam (голям) velik (велик)	velik	veliký	stor	groot	iso kookas suuri
zakon (закон)	zakon	zákon	lov	wet	laki
list (лист)	list	list	blad	bladzijde	lehti sivu

	GREEK Graeca	HUNGARIAN Hungarica	ITALIAN Italica	LATIN Latina	NORWEGIAN Norvegica
170	instituton (ἰνστιτοῦτον) katastema (κατάστημα)	intézet	istituto	institutum	anstalt institutt
171	eisago (εἰσάγω) prosago (προσάγω)	bevezet	introdurre	introducere	innføre innlede
172	eisagoge (εἰσαγωγή) prolegomena (προλεγόμενα) prologos (πρόλογος)	bevezetés	introduzione	introductio praefatio	innførelse innføring innledning introduksjon
173	kineta (κινητά) apographe (ἀπογραφή)	készlet állomány leltár	inventario	inventarium	bestandfortegnelse inventar
174	synergates (συνεργάτης)	társszerző	coautore	co-autor	medforfatter
175	glossa (γλῶσσα) glotta (γλῶττα)	nyelv	lingua	lingua	språk
176	megas (μέγας) megalos (μεγάλος)	nagy	grande	magnus	stor
177	nomos (νόμος)	törvény	legge	lex	lov
178	phyllon (φύλλον)	levél papírlap	foglio	folium	blad

POLISH Polonica	PORTUGUESE Portugallica	RUMANIAN Rumenica	SERBIAN Servica	SLOVAK Slovaca	SWEDISH Suecica
instytut	instituto	institut	zavod (завод) institut (институт)	ústav inštitút	institut anstalt
napisać wstęp wprowadzić	introduzir	introduce	uvoditi (уводити)	uviesť	inleda
słowo wstępne wstęp	introdução	introducere	uvod (увод)	úvod	inledning
inwentarz	inventário	inventar	inventar (инвентар)	inventár	inventarium
współautor	co-autor	coautor	suautor (суаутор)	spoluautor	medförfattare
język	lingua linguagem	limbă limbaj	jezik (језик)	jazyk	språk
wielki duży	grande	mare	velik (велик)	veľký	stor
ustawa prawo	lei	lege	zakon (закон)	zákon	lag
liść	página	foaie filă	list (лист)	list	blad

	ENGLISH Anglica	FRENCH Gallica	GERMAN Germanica	RUSSIAN Russica	SPANISH Hispanica
179	lecture	conférence lecture	Vorlesung Vortrag	doklad (доклад) lekcija (лекция)	lectura lección
	left-hand page →verso				
180	lending (circulation)	prêt	Ausleihe	abonement (абонемент) vydača knig na dom (выдача книг на дом)	préstamo
181	letter (epistle)	lettre	Brief Epistel	pismo (письмо)	carta epístola
	letters →literature				
	lexicon →dictionary				
182	librarian	bibliothécaire	Bibliothekar	bibliotekar' (библиотекарь)	bibliotecario
183	library	bibliothèque	Bibliothek	biblioteka (библиотека)	biblioteca
184	list (register)	liste registre	Verzeichnis	spisok (список) perečen (перечень) ukazateľ (указатель)	lista registro
185	literary remains pl. (remains)	legs pl.	Nachlaß	literaturnoe nasledie (литературное наследие) posmertnye proizvedenija (посмертные произведения)	obras póstumas pl. resto
186	literature (letters pl.)	littérature lettres pl.	Literatur Schrifttum	literatura (литература)	literatura
187	lithograph(y)	lithographie	Lithographie Steindruck	litografija (литография) litografskaja pečať (литографская печать)	litografía

90

BULGARIAN Bulgarica	CROATIAN Croatica	CZECH Bohemica	DANISH Danica	DUTCH Hollandica	FINNISH Fennica
skazka (сказка) doklad (доклад) četene (четене) lekcija (лекция)	predavanje referat lekcija	přednes referát přednáška	foredrag forelæsning	voordacht lezing	esitelmä
knigozaemane (книгозаемане)	posudba zajam	výpůjčka	udlån	uitlening	laina(us)
pismo (писмо)	pismo list	dopis psaní	brev epistel	brief	kirje epistola
bibliotekar (библиотекар)	bibliotekar	knihovník	bibliotekar	bibliothecaris	kirjastonhoitaja
biblioteka (библиотека)	biblioteka knjižnica	knihovna	bibliotek	bibliotheek	kirjasto
spisâk (списък)	popis spisak	seznam	liste fortegnelse	lijst tabel	lista luettelo
literaturno nasledstvo (литературно неследство) posmârtni proizvedenija (посмъртни произведения) *pl.*	nasljedstvo ostavština	(literární) odkaz	bo efterladenskab	nalatenschap	jäännös
literatura (литература) knižnina (книжнина)	književnost literatura	literatura písemnictví	litteratur	literatuur letterkunde	(kauno) kirjallisuus
litografija (литография)	kamenotisak litografija	kamenotisk litografie	litografi stentryk	lithografie steendruk	kivipaiono(s)

		GREEK Graeca	HUNGARIAN Hungarica	ITALIAN Italica	LATIN Latina	NORWEGIAN Norvegica
179		dialexis *(διάλεξις)* paradosis *(παράδοσις)*	előadás felolvasás	discorso lezione lettura	dictio explicatio pertractatio praelectio	forelesning opplese
180		daneismos *(δανεισμός)*	kölcsönzés	prestito	mutuatio librorum	utlån
181		epistole *(ἐπιστολή)*	levél	lettera	epistula litterae *pl.*	brev epistel
182		bibliothekarios *(βιβλιοθηκάριος)*	könyvtáros	bibliotecario	bibliothecarius	bibliotekar
183		bibliotheke *(βιβλιοθήκη)*	könyvtár	biblioteca	bibliotheca	bibliotek
184		elenchos *(ἔλεγχος)* katalogos *(κατάλογος)*	jegyzék lajstrom lista	elenco lista registro	elenchus tabulae *pl.* register index	fortegnelse liste
185		kleronomia *(κληρονομία)* leimma *(λεῖμμα)*	irodalmi hagyaték/örökség hátrahagyott művek *pl.*	lascito	opera relicta (omnia) opera postuma *pl.*	dødsbo etterlatte skrifter *pl.* legat
186		logotechnia *(λογοτεχνία)* philologia *(φιλολογία)*	irodalom	letteratura	litterae *pl.* litteratura	litteratur
187		lithographia *(λιθογραφία)*	kőnyomat	litografia	lithographia opus lithographicum	litografi steintrykk

POLISH Polonica	PORTUGUESE Portugallica	RUMANIAN Rumenica	SERBIAN Servica	SLOVAK Slovaca	SWEDISH Suecica
odczyt prelekcja wykład	conferência	conferinţă prelegere	lekcija (лекција) predavanje (предавање) referat (реферат)	prednes referát prednáška	föredrag
wypożyczanie	empréstimo	împrumutare	pozajmica (позајмица)	vypožičanie	lån
pismo list	carta epístola	scrisoare epistolă	pismo (писмо) list (лист)	list písmo	brev epistel
bibliotekarz	bibliotecário	bibliotecar	bibliotekar (библиотекар)	knihovník	bibliotekarie
biblioteka księgozbór	biblioteca	bibliotecă	biblioteka (библиотека) knjižnica (књижница)	knižnica	bibliotek
lista skorowidz spis	lista	listă	popis (попис) spisak (списак)	súpis zoznam	lista förteckning register
spuścizna	obras póstumas *pl.*	moştenire opere postume *pl.*	nasledstvo (наследство) zaostavština (заоставштина)	(literárny) odkaz	återstod lämningar
literatura	literatura	literatură	književnost (књижевност) literatura (литература)	literatúra písomníctvo	literatur
litografia	litografia	litografie	kamenotisak (каменотисак) litografija (литографија)	kameňotlač litografia	littografi

		ENGLISH Anglica	**FRENCH** Gallica	**GERMAN** Germanica	**RUSSIAN** Russica	**SPANISH** Hispanica
188		little (small)	petit	klein	malen'kij (маленький) nebolšoj (небольшой)	pequeño
189		location mark (shelf mark/number call number)	cote	Stand(ort)nummer Ordnungsnummer Standortsignatur Lokalsignatur	šifr (шифр) poločnyj indeks (полочный индекс) indeks (индекс)	signatura topográfica
190		loose-leaf book (loose-leaf volume loose-leaf)	publication à feuillets mobiles	Loseblattbuch Loseblattsammlung	listkovoe izdanie (листковое издание)	libro de hojas cambiables
		magazine → periodical[2]				
		main → chief				
191		main card	fiche principale	Hauptzettel	osnovnaja kartočka (основная карточка)	ficha principal
		manual → handbook, textbook				
192		manuscript	manuscrit	Handschrift Manuskript	rukopis' (рукопись) manuskript (манускрипт)	manuscrito
193		map	carte géographique	Karte Landkarte	karta (карта)	mapa geográfico
194		margin	marge	Rand Blattrand	pole (поле)	margen
195		matter	composition	Satz	nabor (набор)	composición
		meeting → assembly, sitting				
196		memoir (memorial memorandum)	mémoire mémorial	Denkschrift Memoire Memorandum	memorandum (меморандум)	memorándum memoria(s *pl.*)

BULGARIAN Bulgarica	CROATIAN Croatica	CZECH Bohemica	DANISH Danica	DUTCH Hollandica	FINNISH Fennica
malâk (малък)	mali	malý	lille små	klein	pieni pikku
signatura (сигнатура)	mjesna signatura	signatura knihy	pladssignatur	magazijnnummer boeknummer	paikanmerkki paikkamerkki (hylly)signumi
listovo izdanie (листово издание)	knjiga sastavljena od slobodnih listova	kniha s volnými listy	løsbladsbog	losbladig boek	irtolehtikirja
osnovna kartička (основна картичка) osnoven fiš (основен фиш)	glavni kataloški listić	hlavní záznam	hovedkort hovedseddel	hoofdkaart	pääkortti
râkopis (ръкопис)	rukopis	rukopis	håndskrift manuskript	handschrift manuscript	käsikirjoitus
geografska karta (географска карта)	zemljopisna karta	(zeměpisná) mapa	landkort	geografische kaart kaart landkaart	kartta
pole na kniga (поле на книга)	rub (stranice) margina	okraj stránky	margin	kantlijn marge	vierus marginaali
nabor (набор)	slog	sazba	sats	zetsel	lados
memorandum (меморандум)	spomenica memorandum	pamětní spis	mindeskrift	gedenkschrift	muistokirjoitus

	GREEK Graeca	HUNGARIAN Hungarica	ITALIAN Italica	LATIN Latina	NORWEGIAN Norvegica
188	mikros (μικρός)	kis	piccolo	parvus exiguus	lille liten små *pl.*
189	arithmos apothekes (ἀριθμός ἀποθήκης) episema (ἐπίσημα)	helyrendi szám raktári jelzet	segnatura collocazione	signatura loci	(plas)signatur boksignatur
190	astachoton (ἀστάχωτον)	szabadlapos könyv könyv cserélhető lapokkal	libro a pagine mobili	liber cum foliis mobilibus	løsbladbok
191	kyrion deltion (κύριον δελτίον)	főlap	scheda principale	charta/scida principalis	hovedkort hovedseddel
192	cheirographon (χειρόγραφον) autographon (αὐτόγραφον)	kézirat	manoscritto originale	autographum chirographum manuscriptum	håndskrift manuskrift
193	(geografikos) chartes (γεωγραφικός χάρτης)	térkép	carta geografica	charta	kart landkart
194	perithorion (περιθώριον)	margó lapszél	margine	margo	marg
195	stoicheiothesia (στοιχειοθεσία)	szedés	composizione	compostitio typorum	sats
196	hypomnema (ὑπόμνημα)	emlékirat	memoriale memorandum	in memoriam ... libellus commentarius memorandum	memorandum minneskrift

POLISH Polonica	PORTUGUESE Portugallica	RUMANIAN Rumenica	SERBIAN Servica	SLOVAK Slovaca	SWEDISH Suecica
mały	pequeno	mic	mali (мали)	malý	liten små(-)
sygnatura	cota	local numǎr numǎr de magazie	signatura (сигнатура)	signatúra	lokalsignum
wydawnictwo skoroszytowe/ luźnokartkowe	livro de folhas soltas	carte cu foi libere	knjiga sa slobodnim listovima (књига са слободним листовима)	kniha s voľnými listami	lösbladsbok
karta główna	ficha principal	fişă principală	glavni kataloški listić (главни каталош- ки листић)	hlavný lístok	huvudkort
manuskrypt rękopis	manuscrito	manuscris original	rukopis (рукопис)	rukopis	handskrift
mapa	mapa geográfico	hartă	geografska karta (географска карта)	(zemepisná) mapa	karta
margines	margem	margine	margina (маргина)	okraj	marginal
skład	composição	cules zaţ	slog (слог)	sadzba	sättning
memorial pamiętnik	memorandum nota	memorial memorandum	spomenica (споменица) memorandum (меморандум)	pamätný spis	minnesskrift

	ENGLISH Anglica	FRENCH Gallica	GERMAN Germanica	RUSSIAN Russica	SPANISH Hispanica
197	memoirs *pl.*	mémoires *pl.*	Memoiren Erinnerungen *pl.*	memuary (мемуары) *pl.* vospominanija (воспоминания) *pl.*	memorias *pl.*
	memorandum →memoir				
	memorial →memoir				
198	memorial speech	discours commé- moratif	Gedächtnisrede	nadgrobnaja reč' (надгробная речь)	discurso conme- morativo
199	memorial volume (festschrift)	hommage à... mélanges *pl.* offerts à...	Festschrift	jubilejnyj sbornik (юбилейный сборник)	en honor de... volumen conmemora- tivo homenaje a...
200	microfilm	microfilm	Mikrofilm	mikrofilm (микрофильм) mikroplenka (микропленка)	microfilme
201	miniature	miniature	Miniatur	miniatjura (миниатюра)	miniatura
202	minutes *pl.* (record[2] report[2])	procès-verbal	Protokoll	protokol (протокол)	acta de la sesión
203	miscellanea *pl.* (miscellany miscellaneous writings)	miscellanea mélanges *pl.* miscellanées *pl.*	Miszellen *pl.* Miszellaneen *pl.*	raznoe (разное)	miscelánea *pl.*
	miscellaneous writings →miscellanea				
	miscellany →miscellanea				
	misprint →printer's error				
204	monograph(y)	monographie	Monographie	monografija (монография)	monografía

BULGARIAN Bulgarica	CROATIAN Croatica	CZECH Bohemica	DANISH Danica	DUTCH Hollandica	FINNISH Fennica
memoari (мемоари) *pl.* spomeni (спомени) *pl.*	memoari *pl.*	paměti *pl.*	memoirer *pl.* erindringer *pl.*	memoires *pl.* herinneringen *pl.*	memoaarit *pl.* muistelmat *pl.*
vâzpomenatelna reč (възпоменателна реч)	posmrtni govor spomen-slovo	chvalořeč připomínková řeč	mindetale	lijkrede	ruumissaarna
jubileen sbornik (юбилеен сборник)	spomenica spomen-knjiga	památník slavnostní vydání	mindeskrift	feestbundel	juhlakirja muistojulkaisu
mikrofilm (микрофилм)	mikrofilm	mikrofilm	mikrofilm	mikrofilm	mikrofilmi
miniatjura (миниатюра)	minijatura	miniatura	miniatur	miniatuur	pienoiskuva pienoismaalaus
protokol (протокол)	zapisnik protokol	protokol	forhandlinger protokol	verslag protocol	pöytäkirja
razni (разни) *pl.*	različit miscelanea *pl.*	smíšený smíšenina	blandet blandinger *pl.*	verspreide opstellen gemengd	sekalasia
monografija (монография)	monografija	monografie	monografi	monografie	monografia

	GREEK Graeca	HUNGARIAN Hungarica	ITALIAN Italica	LATIN Latina	NORWEGIAN Norvegica
197	apomnemoneumata (ἀπομνημονεύματα) pl.	emlékezések pl.	memorie pl. ricordi pl.	memoriae pl. commemorationes pl.	erindringer pl. memoarer pl.
198	logos epitaphios (λόγος ἐπιτάφιος) epikedeios logos (ἐπικήδειος λόγος)	emlékbeszéd	commemorazione necrologia	laudatio (funebris) commemoratio	minnetale

199	iobilaia sylloge
	(ἰωβιλαία συλλογή)
		emlékkönyv	pubblicazione
			  commemorativa
			in onore . . .
				liber memorialis
				  ad honorem . . .
				libellus commenta-
				rius in memoriam . . .
					minnebok
					minneskrift

200	mikrophilm (μικροφίλμ)	mikrofilm	microfilm	taeniola photographi- ca microfilm	mikrofilm
201	mikrographia (μικρογραφία) miniatura (μινιατούρα)	miniatúra	miniatura	miniatura	miniatyr
202	praktika (πρακτικά) pl.	jegyzőkönyv	processo verbale protocollo	tabulae protocollum regestrum	protokoll
203	syngrammata mikta (συγγράμματα μικτά) pl.	vegyes (cikkek pl.)	diverso miscellanea pl.	mixtus miscellanea pl.	blandet
204	monografia (μονογραφία)	monográfia	monografia	monographia	monografi

100

POLISH Polonica	PORTUGUESE Portugallica	RUMANIAN Rumenica	SERBIAN Servica	SLOVAK Slovaca	SWEDISH Suecica
wspomnienie	memórias *pl.*	memorii *pl.* amintiri *pl.*	memoari (мемоари) *pl.*	rozpomienky *pl.*	memoarer *pl.*
mowa ku uczcze- niu pamięci	necrológio	cuvîntare come- morativă	posmrtni govor (посмртни говор) spomen-slovo (спомен-слово)	reč na pamiatku niekoho	minnestal
księga pamiąt- kowa memoriał wydawnictwo jubileuszowe	publicação comemorativa	ediție festivă/ /omagială omagiu	spomenica (споменица) spomen-knjiga (спомен-књига)	pamätná kniha pamätnica	festskrift
mikrofilm	microfilme	microfilm	mikrofilm (микрофилм)	mikrofilm	mikrofilm
miniatura	miniatura	miniatură	minijatura (минијатура)	drobnomaľba miniatúra	miniatyr
protokół	protocolo	proces-verbal protocol	zapisnik (записник) protokol (протокол)	zápisnica protokol	protokoll förhandlingar *pl.*
mieszany miscellanea *pl.*	miscelânea vária	amestecat divers miscelanee *pl.*	mešovit (мешовит) različit (различит)	zmiešaný rôzne rozličný	blandad blandade skrifter *pl.* allehanda
monografia	monografia	monografie	monografija (монографија)	monografia	monografi

	ENGLISH Anglica	FRENCH Gallica	GERMAN Germanica	RUSSIAN Russica	SPANISH Hispanica
	multilingual →**polyglot**				
	municipal library →**city library**				
205	**music**	musique	Musik	muzyka (музыка)	música
206	**name**	nom	Name	imja (имя) familija (фамилия)	nombre apellido
207	**national library**	bibliothèque nationale	Staatsbibliothek	nacionaľnaja biblioteka (национальная библиотека)	biblioteca nacional
	necrology →**obituary notice**				
208	**new** **(recent)**	neuf récent	neu jüngst	novyj (новый)	nuevo reciente
209	**newspaper** **(journal** **gazette)**	journal gazette	Zeitung	gazeta (газета) ežednevnik (ежедневник)	diario gaceta
210	**no place no date** **n.p.n.d.**	sans lieu ni date s.l. n. d.	ohne Orts-und Jahresangabe o.O.u.J.	bez ukazanija mesta i goda b. m. i g. (без указания места и года б. м. и г.) bez mesta, bez goda b. m., b. g. (без места, без года б. м., б. г.)	sin lugar sin año s.l.s.a.
211	**notation**	indice de classification	Notation Bezeichnungsweise	indeksacija (индексация)	notación
212	**note[1]** **(comment[2])**	remarque note	Bemerkung Anmerkung Notiz Note	zamečanie (замечание) zametka (заметка) primečanie (примечание)	advertencia apuntación nota

BULGARIAN Bulgarica	CROATIAN Croatica	CZECH Bohemica	DANISH Danica	DUTCH Hollandica	FINNISH Fennica
muzika (музика)	muzika glazba	hudba	musik	muziek	musiikki
ime (име)	ime	jméno	navn	naam	nimi
narodna/ nacionalna biblioteka (народна/ национална библиотека)	narodna knijižnica	národní knihovna	nationalbibliotek	nationale bibliotheek	kansalliskirjasto
nov (нов)	nov	nový	ny	nieuw	uusi
vestnik (вестник) ežednevnik (ежедневник)	list novine	noviny *pl.* deník	avis blad dagblad	dagblad krant nieuwsblad	sanomalehti
bez mjasto i godina b. m. i g. (без място и година б. м. и г.)	bez mjesta i godine bez mj. i god.	bez místa a roku b. m. r.	uden sted og år u.s.o.å.	zonder plaats en jaar z.pl.e.j.	ilman painopaik- kaa ja vuotta i.p.& v.
indeksacija (индексация)	označavanje stručno obilježa- vanje	notace	notation	notatie	merkkijärjestelmä merkistö
beležka (бележка) zabeležka (забележка)	napomena opaska	poznámka	anmærkning bemærkning	noot opmerking	huomautus muistutus

	GREEK Graeca	HUNGARIAN Hungarica	ITALIAN Italica	LATIN Latina	NORWEGIAN Norvegica
205	musike (μουσική)	zene	musica	musica	musikk
206	onoma (ὄνομα)	név	nome	nomen	navn
207	ethnike bibliotheke (ἐθνική βιβλιοθήκη)	országos könyvtár	biblioteca nazionale	bibliotheca nationalis	nasjonalbibliotek
208	neos (νέος)	új	nuovo recente rinnovato	novus recens	ny
209	ephemeris (ἐφημερίς)	hírlap újság	giornale gazzetta	acta diurna *pl.* diurna *pl.*	avis blad
210	aneu topu kai chronu a.t.k.ch. (ἄνευ τόπου καί χρόνου, α.τ.κ.χ.)	hely és év nélkül h.é.n.	senza luogo senza data s.l.s.d.	sine loco sine anno s.l.s.a.	uten sted uten år u.s.u.å.
211	hyposemeiosis (ὑποσημείωσις)	szakjelzet jelzet	notazione del sistema di classificazione decimale	notatio	notasjon det anvendte signeringssystem
212	scholion (σχόλιον)	megjegyzés jegyzet	osservazione nota	nota	anmerkning bemerkning

POLISH Polonica	PORTUGUESE Portugallica	RUMANIAN Rumenica	SERBIAN Servica	SLOVAK Slovaca	SWEDISH Suecica
muzyka	música	muzică	muzika (музика) glazba (глазба)	hudba	musik
imię nazwa nazwisko	nome	nume	ime (име)	meno	namn

biblioteka
  narodowa
	biblioteca
	  nacional
		bibliotecă
		  naţională
			narodna
			  biblioteka
			  (народна
			    библиотека)
				národná knižnica	nationalbibliotek

nowy	novo	nou	nov (нов)	nový	ny
gazeta dziennik	jornal quotidiano	ziar gazetă jurnal	list (лист) novine (новине) pl.	noviny pl. denník	tidning

bez miejsca bez
  roku
b. m. r.
	sem lugar nem
	  data
	s. l. n. d.
		fără loc şi an
		f. l. ş. a.
			bez mesta i
			  godine
			b. m. i g.
			(без места и
			  године
			  б. м. и г.)
				bez miesta a bez
				  roku
				b. m. b. r.
					utan ort utan år
					u. o. u. å.

znakowanie	notação	notaţie	indeksiranje (индексирање)	notácia	anmärkning
przypis uwaga	nota observação	însemnare observaţie	napomena (напомена) opaska (опаска)	poznámka	anmärkning

	ENGLISH Anglica	FRENCH Gallica	GERMAN Germanica	RUSSIAN Russica	SPANISH Hispanica
	note² →annotation				
213	novel	roman	Roman	roman (роман) povest' (повесть)	novela
	n.p.n.d. →no place no date novelette →short story				
214	number¹ (figure)	nombre numéro chiffre	Zahl	nomer (номер) cifra (цифра)	número cifra
215	number² (copy²)	cahier brochure	Heft Nummer Lieferung	nomer (номер)	fascículo
	nuptial song →bridal song				
	obituary →obituary notice				
216	obituary notice (obituary necrology)	notice nécrologique obituaire necrologie	Nekrolog Nachruf	nekrolog (некролог)	necrología
217	oblong	oblong	quer	poperečnyj (поперечный)	oblongo
218	of	d' de	von aus über	iz (из)	de, desde
219	office	office bureau	Amt Büro	upravlenie (управление)	oficina
220	official	officiel	amtlich offiziell	oficialnyj (официальный)	oficial
221	offprint (reprint¹ separate)	tirage à part tiré à part	Sonderdruck Sonderabdruck Separatabdruck	otdelnyj ottisk (отдельный оттиск)	impreso por separado tirada/tiraje aparte

BULGARIAN Bulgarica	CROATIAN Croatica	CZECH Bohemica	DANISH Danica	DUTCH Hollandica	FINNISH Fennica
roman (роман) povest (повест)	roman	román	roman	roman	romaani
cifra (цифра) nomer (номер)	broj	číslo	tal antal nummer	getal aantal nummer	luku numero laskumerkki
knižka (книжка) svezka (свезка) broj (брой)	sveska broj	sešit číslo	hæfte nummer	heft	numero
nekrolog (некролог) skrâbna vest (скръбна вест)	nekrolog	nekrolog	nekrolog	necrologie	muistokirjoitus nekrologi
prodâlgovat (продълговат)	duguljast	podlouhlý	aflang i tværformat	oblong langwerpig	pitkulainen soikea
na (на) ot (от) za (за); vârhu (върху)	od iz	od	af	uit van	-n
služba (служба) upravlenie (управление)	služba ured	úřad služba	embede kontor	ambt	virasto virka
oficialen (официален)	zvaničan služben	úřední	officiel	officiëel ambtelijk	virallinen
otdelen otpečatâk (отделен отпечатък)	poseban/zaseban otisak separat	zvláštní otisk separát	særtryk	overdruk	ylipainos eripainos

	GREEK Graeca	HUNGARIAN Hungarica	ITALIAN Italica	LATIN Latina	NORWEGIAN Norvegica
213	mythistorema (μυθιστόρημα)	regény	romanzo	fabula Romanensis	roman
214	arithmos (ἀριθμός)	szám	numero	numerus	nummer tall
215	numero (νούμερο) phyllon (φύλλον)	füzet	fascicolo	fasciculus	hefte levering nummer
216	nekrologion (νεκρολόγιον)	nekrológ	necrologia	commemoratio	nekrolog
217	epimekes (ἐπιμήκης) enkarsios (ἐγκάρσιος)	haránt fekvő	oblungo	oblongus	avlang langaktig
218	apo (ἀπό) ek, ex (ἐκ, ἐξ)	-nak, -nek -tól, -től; -ból, -ből -ról, -ről	da di	ab, a ex, e de	av fra
219	hyperesia (ὑπηρεσία)	hivatal	ufficio	magistratus	embete kontor
220	episemos (ἐπίσημος)	hivatalos	ufficiale	officialis publicus	embets- offisiell
221	anatypoma (ἀνατύπωμα) ektaktos anatyposis (ἔκτακτος ἀνατύπωσις)	különlenyomat	tiratura a parte estratto	impressum separatum separatum	separatavtrykk

POLISH Polonica	PORTUGUESE Portugallica	RUMANIAN Rumenica	SERBIAN Servica	SLOVAK Slovaca	SWEDISH Suecica
romans powieść	romance	roman	roman (роман)	román	roman
liczba numer	número	număr cifră	broj (број)	číslo cifra	tal nummer
zeszyt	fasciculo	fasciculă	sveska (свеска) broj (број)	zošit	häfte nummer
nekrolog	necrológio	necrolog	nekrolog (некролог)	nekrológ	nekrolog
podłużny poprzeczny	oblongo	oblong transverzal	duguljast (дугуљаст)	podlhovastý pozdĺžny	avlång i tvärformat
od z(e)	do, da; dos, das de por	de din	od (од) iz (из)	od z zo	av om
urząd	escritório ofício	oficiu	ured (уред)	úrad	ämbete kontor tjänst
oficjalny urzędowy	oficial	oficial	zvaničan (званичан) služben (службен)	oficiálny úradný	officiell
odbitka autorska nadbitka	separata tiragem à parte	extras tragere separată	poseban/zaseban otisak (посебан/засебан отисак) separat (сепарат)	zvláštny/autorský výtlačok separát	särtryck

	ENGLISH Anglica	FRENCH Gallica	GERMAN Germanica	RUSSIAN Russica	SPANISH Hispanica
	old →ancient				
222	**on**	sur de	über um von auf	na (на) po (по) o(b) (об)	sobre de
	on the basis of →based/founded on the				
223	**only** *a.* **(sole unique)**	unique	einzig	edinstvennyj (единственный) unikaĺnyj (уникальный)	único
	oration →speech				
224	**order**[1] **(decree)**	décret ordre	Verordnung	ukaz (указ) dekret (декрет) postanovlenie (постановление)	orden decreto
225	**order**[2]	commande	Bestellung	zakaz (заказ)	pedido
	order[3] *v.* →arrange				
226	**original**	original	original ursprünglich	originaĺnyj (оригинальный) podlinnyj (подлинный)	original
227	**original edition (first edition)**	édition originale/ /première	Originalausgabe Erstausgabe	pervoe izdanie (первое издание)	edición original/ /príncipe
228	**page**	page	Seite	stranica (страница)	página
	pagination →paging				
229	**paging (pagination)**	pagination	Paginierung	numeracija stranic (нумерация страниц) paginacija (пагинация)	paginación

BULGARIAN Bulgarica	CROATIAN Croatica	CZECH Bohemica	DANISH Danica	DUTCH Hollandica	FINNISH Fennica
na (на) po (по) vârhu (върху) za (за) vâz (въз)	o po vrh na	o ob nad na	om over på	over van op	-sta -stä
edinstven (единствен) unikalen (уникален)	jedini	jediný	enestående eneste	enig	ainoa
naredba (наредба) postanovlenie (постановление) ukaz (указ)	naredba uredba	nařízení ustanovení výnos	forordning ordre	order verordening	määräys säädös
porâčka (поръчка)	narudžbina narudžba	objednávka	bestilling	bestelling	tilaus
originalen (оригинален	originalan	původní	original	origineel	alkuperäinen
pârvo izdanie (първо издание)	prvo izdanje	první edice/vydání	originaludgave	eerste uitgave	ensipainos kantapainos
stranica (страница)	stranica	stránka	side (i bog)	bladzijde pagina	sivu
paginacija (пагинация)	paginacija	stránkování paginace	paginering	paginering nummering van der bladzijden	sivunumerointi

	GREEK Graeca	HUNGARIAN Hungarica	ITALIAN Italica	LATIN Latina	NORWEGIAN Norvegica
222	se *(σέ)* gia *(γιά)*	-on, -en, -ön -ról, -ről -ra, -re	da su sopra	de supra super	for på
223	monos *(μόνὸς)*	egyetlen	unico	unicus unus solus	eneste enestående
224	diatagma *(διάταγμα)*	rendelet	decreto ordinamento ordinanza	decretum edictum praescriptum	forordning
225	parangelia *(παραγγελία)*	(meg)rendelés	ordinazione	mandatio librorum	bestilling
226	prototypos *(πρωτότυπος)*	eredeti	originale	originalis	opprinnelig original
227	ekdosis prote *(ἔκδοσις πρώτη)*	eredeti kiadás első kiadás	edizione originale/principe	editio princeps/ originalis	første utgave
228	selis *(σελίς)* selida *(σελίδα)*	oldal	pagina	pagina	side
229	selidosis *(σελιδώσις)*	lapszámozás oldalszámozás	paginazione	paginatio	paginering

112

POLISH Polonica	PORTUGUESE Portugallica	RUMANIAN Rumenica	SERBIAN Servica	SLOVAK Slovaca	SWEDISH Suecica
o z(e) na	a por para sobre, sôbre	asupra despre la	o (о) po (по) vrh (врх) na (на)	o nad na	om över på
jedyny	único	unic	jedini (једини)	jediný	enda
dekret porządek rozporządzenie	decreto ordem	decret ordin	naredba (наредба) uredba (уредба)	nariadenie príkaz	förordning
zamówienie	encomenda pedido	comandă	narudžbina (наруџбина)	objednávka	beställning
oryginalny	original	original	originalan (оригиналан)	pôvodný	original
pierwodruk pierwsze wydanie	primeira edição	ediţie princeps prima ediţie	prvo izdanje (прво издање)	prvé vydanie	första upplaga originalupplaga
strona	página	pagină	stranica (страница)	strana	sida boksida
liczbowanie paginacja	paginação	numărătoarea paginilor paginaţie	paginacija (пагинација)	stránkovanie paginácia	paginering

8

	ENGLISH Anglica	FRENCH Gallica	GERMAN Germanica	RUSSIAN Russica	SPANISH Hispanica
230	panegyric	panégyrique éloge	Lobgesang Preislied	chvalebnyj gimn (хвалебный гимн) panegirik (панегирик)	panegírico elogio
231	paper[1] paper[2] →study	papier	Papier	bumaga (бумага)	papel

232	paperback
	(paper-bound book)
		livre broché	geheftetes/broschiertes
			  Buch
			Paperback
				kniga v bumažnom
				  perepl̈ete
				(книга в бумажном
				  переплёте)
					libro de rústica

233	papyrus	papyrus	Papyrus	papirus (папирус)	papiro
234	paragraph	paragraphe	Paragraph	paragraf (параграф) abzac (абзац)	párrafo
235	parchment	parchemin	Pergament	pergament (пергамент)	pergamino
236	part	partie part	Teil	časť (часть)	parte

237	passim	passim	passim	vezde (везде)
				v raznych
				  mestach
				(в разных
				  местах)
					en diversos sitios

	pen-name →pseudonym				
238	periodical[1]	périodique[1]	periodisch	periodičnyj (периодичный) periodiceškij (периодический)	periódico
239	periodical[2] (journal magazine)	périodique[2] magazine	Zeitschrift	periodičeskoe izdanie (периодическое издание) žurnal (журнал) periodika (периодика)	revista periódico
	photo →photograph				

BULGARIAN Bulgarica	CROATIAN Croatica	CZECH Bohemica	DANISH Danica	DUTCH Hollandica	FINNISH Fennica
hvalebna pesen (хвалебна песен) pohvalno slovo (похвално слово)	pohvalna pjesma	chvalozpěv	lovsang	lofzang	ylistely ylistyslaulu
hartija (хартия)	papir	papír	papir	papier	paperi
izdanie s meka podvârzija (издание с мека подвързия)	broširana knjiga brošura	brožovaná kniha	hæftet bok	gebrocheerd boek niet gebonden boek paper back	nidottu kirja
papirus (папирус)	papirus	papyrus	papyrus	papyrus	papyrus
paragraf (параграф)	abzac paragraf	oddíl paragraf	paragraf	paragraaf	pykälä paragrafi
pergament (пергамент)	pergament	pergamen	pergament	perkament	pergamentti
čast (част) djal (дял)	dio	část díl	del	deel gedeelte	osa
tuk-tam (тук-там)	ovdje-ondje razbacano	passim	passim på forskellige steder	passim	hajakohdin usseissa kohdin
periodičen (периодичен)	periodičan	periodický	periodisk	periodiek	periodinen
spisanie (списание)	časopis revija	časopis revue	tidsskrift	tijdschrift periodiek	aikakauskirja aikakauslehti

	GREEK Graeca	HUNGARIAN Hungarica	ITALIAN Italica	LATIN Latina	NORWEGIAN Norvegica
230	penegyrikos (logos) (πανηγυρικός λόγος)	dicsőítő ének dicshimnusz	inno elogio	elogia laudatio paean	lovord lovtale
231	chartes (χάρτης)	papír	carta	charta	papir
232	chartodeton biblion (χαρτοδέτον βιβλίον)	fűzött könyv papírfedelű zsebkönyv	,,brochure'' brossura rilegatura alla rustica	liber vilis in tegimento/ tegumine chartaceo	uinnbundet (bok) heftet (bok) billigbok
223	papyros (πάπυρος)	papírusz	papiro	papyrus	papyrus
234	paragraphos (παράγραφος)	szakasz	paragrafo articolo	paragraphus articulus caput	avsnitt paragraf
235	pergamene (περγαμηνή)	pergamen	pergamena	charta pergamena	pergament
236	meros (μέρος)	rész	parte	pars	del
237	hekastachu (ἑκασταχοῦ) pantachu (πανταχοῦ)	passzim több helyütt	passim	passim	passim på forskjellige steder pl.
238	periodikos (περιοδικός)	időszaki	periodico	periodicus	periodisk
239	periodikon (περιοδικόν)	folyóirat	periodico rivista	commentarius periodicum ephemeris	periodika periodisk skrift tidsskrift

POLISH Polonica	PORTUGUESE Portugallica	RUMANIAN Rumenica	SERBIAN Servica	SLOVAK Slovaca	SWEDISH Suecica
pieśń pochwalna panegiryk	elogio panegírico	elogiu panegiric	pohvalna pesma (похвална песма)	oslavná báseň	lovkväde lovsång
papier	papel	hîrtie	papir (папир)	papier	papper
książka nie- oprawna	livro brochado	carte broşată	broširana knjiga (broширана књига)	brožovaná kniha brožúra	häftad bok
papyrus	papiro	papirus	papirus (папирус)	papyrus	papyrus
rozdział paragraf ustęp	parágrafo	paragraf	abzac (абзац) paragraf (параграф)	oddiel paragraf	avdelning paragraf
pergamin	pergaminho	pergament	pergament (пергамент	pergamen	pergament
część dział	parte	parte	deo (део)	časť diel	del stycke
passim	a cada passo	passim	ovde-onde (овде-онде) razbacano (разбацано)	tu i tam passim roztratene	här och där passim på spridda ställen
periodyczny	periódico	periodic	periodičan (периодичан)	periodický	periodisk
czasopismo	periódico revista	revistă	časopis (часопис) revija (ревија)	časopis	periodisk skrift tidskrift

	ANGLICA Anglica	FRENCH Gallica	GERMAN Germanica	RUSSIAN Russica	SPANISH Hispanica
240	**photograph** **(photo)**	photographie	Photographie	foto(grafija) (фотография) fotosnimok (фотоснимок)	fotografía
	phototype **→collotype**				
241	**picture** **(figure** **illustration)**	image figure	Bild Abbildung Figur	kartinka (картинка) illjustracija (иллюстрация) risunok (рисунок)	imagen ilustración figura
242	**piece**	pièce	Stück	štuka (штука)	pieza
243	**place**	lieu	Ort	mesto (место)	lugar
244	**plagiarism**	plagiat	Plagiat	plagiat (плагиат)	plagio
245	**plate**	planche	Tafel Bildtafel	tablica (таблица) vkladnoj list (вкладной лист)	lámina plancha
246	**play** **(drama)**	pièce (de théâtre) drame	Schauspiel Bühnenwerk Drama	p'esa (пьеса) drama (драма)	pieza de teatro drama
247	**pocket-book** **(pocket-companion)**	édition de poche	Taschenbuch	karmannyj spravočnik (карманный справочник) rukovodstvo (руководство)	libreta librito
248	**poem** **(verse)**	vers poème	Dichtung	stich (стих) stichotvorenie (стихотворение)	verso poema
249	**poet**	poète	Dichter	poèt (поэт)	poeta
250	**poetry**	poésie	Poesie	poèzija (поэзия)	poesía

BULGARIAN Bulgarica	CROATIAN Croatica	CZECH Bohemica	DANISH Danica	DUTCH Hollandica	FINNISH Fennica
foto(grafija) (фотография) snimka (снимка)	fotografija	fotografie	fotografi	fotografie	valokuva
kartina (картина) risunka (рисунка) obraz (образ) figura (фигура)	ilustracija slika	obraz vyobrazení ilustrace	billede illustration figur afbildning	illustratie afbeelding	illustraatio kuva
parče (парче) broj(ka) (бройка)	komad parče	kus	stykke	stuk	kappale
mjasto (място)	mjesto	místo	sted	plaats	paikka
plagiat (плагиат)	plagijat književna krada	plagiát	plagiat	plagiaat	kirjallinen varkaus plagiaatti
tablica (таблица)	tablica tabela	tabule	planche tavle	plaat tafel	taulu
piesa (пиеса) drama (драма)	kazališni komad drama	činohra drama	drama skuespil	toneelstuk drama	draama näytelmä

džoben
  spravočnik
(джобен
  справочник)
	priručnik	příručka
		rukověť
			håndbog	naslagwerk
				handleiding
					taskukirja

stihotvorenie (стихотворение)	stih pjesma	báseň	digt digtning	gedicht vers	runo runoelma runous
poet (поет)	pjesnik	básník	digter	dichter	runoilija
poezija (поезия)	poezija pjesništvo	básnictví poezie	poesi	poëzie	runous

	GREEK Graeca	HUNGARIAN Hungarica	ITALIAN Italica	LATIN Latina	NORWEGIAN Norvegica
240	photographia (φωτογραφία)	fénykép	fotografia	photographia	fotografi
241	eikon (εἰκών) eikonographia (εἰκονογραφία)	kép illusztráció ábra	illustrazione immagine figura	illustratio imago figura	avbilde bilde illustrasjon
242	kommati (κομμάτι)	darab	pezzo	pars	stykke
243	topos (τόπος)	hely	luogo	locus	plass sted
244	logoklope (λογοκλοπή) logoklopia (λογοκλοπία)	plágium	plagio	plagium	plagiat plagiering
245	pinax (πίναξ)	képtábla	tavola	tabula	plansje plate tavle
246	theatrikon ergon (θεατρικόν ἔργον) drama (δρᾶμα) theama (θέαμα)	színmű színdarab dráma	dramma	drama fabula theatralis	drama skuespill
247	encheiridion (ἐγχειρίδιον)	zsebkönyv	taccuino	enchiridion compendium vademecum liber forma minore	håndbok lommebok
248	poiema (ποίημα)	költemény	opera poetica poema verso	versus poëma	dikt
249	poietes (ποιητής)	költő	poeta	poëta	dikter
250	poiesis (ποίησις)	költészet	poesia	ars poëtica poësis	dikt diktekunst diktning poesi

POLISH Polonica	PORTUGUESE Portugallica	RUMANIAN Rumenica	SERBIAN Servica	SLOVAK Slovaca	SWEDISH Suecica
fotografia zdjęcie	fotografia	fotografie	fotografija (фотографија)	fotografia	fotografi
ilustracja obraz figura rysunek	figura ilustração imagem	ilustraţie imagine figură	slika (слика) ilustracija (илустрација)	obraz ilustrácia	illustration bild figur
sztuka	peça pedaço	bucată	komad (комад) parče (парче)	kus	stycke
miejsce	lugar	loc	mesto (место)	miesto	ort
plagiat	plágio	plagiat	plagijat (плагијат)	plagiát	plagiat
tablica	tábua	planşă	tablica (таблица) tabela (табела)	tabuľa	plansch
sztuka teatralna utwór drama- tyczny	drama	dramă	pozorišni komad (позоришни комад) drama (драма)	činohra dráma	drama skådespel
poradnik	livro de notas livro portátil	memorator mic manual	priručnik (приручник)	príručka rukoväť	anteckningsbok
wiersz	poema verso	vers poem	stih (стих) pesma (песма)	báseň	dikt
poeta	poeta	poet	pesnik (песник)	básnik	diktare skald författare
poezja	poema poesia	poezie	poezija (поезија) pesništvo (песништво)	básnictvo poézia	poesi skaldekonst

	ENGLISH Anglica	FRENCH Gallica	GERMAN Germanica	RUSSIAN Russica	SPANISH Hispanica
251	**polemic** **(polemical treatise** **controversial** **pamphlet)**	écrit polémique polémique	Polemik Streitschrift	polemičeskaja staťja (полемическая статья)	polémica opúsculo polémico
	polemical treatise →**polemic**				
252	**polyglot** **(multilingual)**	polyglotte multilingue	mehrsprachig polyglott Polyglott	mnogojazyčnyj (многоязычный) poliglot (полиглот)	poligloto
253	**portrait**	portrait	Porträt Bildnis	portret (портрет)	retrato
254	**posthumous**	posthume	postum	posmertnyj (посмертный)	póstumo
255	**postscript**	post-scriptum	Nachschrift	postskriptum (постскриптум) pripiska (приписка)	posdata
256	**preface** **(foreword)**	préface avant-propos propos liminaire	Vorwort Vorrede	predislovie (предисловие)	prefacio
257	**prepare**	préparer	vorbereiten	prigotovljať (приготовлять) podgotavlivať (подготавливать)	preparar
258	**prescription** **(instructions)**	prescription	Vorschrift	predpisanie (предписание)	prescripción reglamento
259	**president** **(chairman)**	président	Präsident	predsedateľ (председатель) prezident (президент)	presidente
260	**press**[1]	presse	Presse	pečať (печать) pressa (пресса)	prensa
	press[2] →**printing office** **principal** →**chief**				

BULGARIAN Bulgarica	CROATIAN Croatica	CZECH Bohemica	DANISH Danica	DUTCH Hollandica	FINNISH Fennica
diskusija (дискусия) polemika (полемика)	polemičan spis	polemický spis polemika	ordstrid strid	dispuut polemiek	poleeminen kirjoitus väittelykirjoitus
mnogoezičen (многоезичен) poliglot (полиглот)	mnogojezičan	několikajazyčný polyglotní	flersproget polyglot	veeltalig	monikielinen polyglotti
portret (портрет)	portret	portrét	portræt	portret	muotokuva
posmârten (посмъртен)	posmrtni	postmrtný	efterladt posthum	postuum	postuumi kuoleman jälkeen julkaistu
poslepis (послепис) pripiska (приписка)	dodatak postskriptum	douška přípisek	efterskrift postskriptum	naschrift	jälkipuhe jälkikirjoitus postskriptumi
predgovor (предговор)	predgovor	předmluva úvod	indledning	voorwoord voorrede	johdanto alkulause
prigotvjam (приготвям) podgotvjam (подготвям)	pripraviti pripremiti	připravit uchystat	forberede	voorbereiden	valmistaa valmistella
predpisanie (предписание)	pravilo propis	předpis pravidlo	regel reglement	voorschrift regel reglement	ohje sääntö
predsedatel (председател)	predsjednik	předseda president	præsident	president voorzitter	presidentti
pečat (печат) presa (преса)	tisak štampa	tisk	pressen	druk pers	paino

	GREEK Graeca	HUNGARIAN Hungarica	ITALIAN Italica	LATIN Latina	NORWEGIAN Norvegica
251	polemike (πολεμική)	vita(irat)	polemica	disputa disputatio	ordstrid polemikk
252	polyglossos (πολύγλωσσος)	többnyelvű	poliglotto	multilinguis polyglottus	flerspråklig polyglot
253	portraito (πορτραῖτο) prosopografia (προσωπογραφία)	arckép	ritratto	imago effigies	portrett
254	metathanatios (μεταθανάτιος)	posztumusz	postumo	posthumus	etterlatt posthum
255	hysterographon (ὑστερόγραφον)	utóirat	poscritto	postscriptum	etterskrift
256	prologos (πρόλογος) prolegomena (προλεγόμενα)	előszó	prefazione	praefatio procemium	forord
257	paraskeuazo (παρασκευάζω) proetoimazo (προετοιμάζω)	előkészít	preparare	parare praeparare	forberede
258	diatage (διαταγή) kanon (κανών) horos (ὅρος)	előíras szabály	regola(mento)	formula praescriptum praeceptum regula	forskrift regel
259	proedros (πρόεδρος)	elnök	presidente	praeses	formann president
260	typos (τύπος) piesterion (πιεστήριον)	sajtó	pressa stampa	prelum typographi- cum acta diurna *pl.*	presse trykk

POLISH Polonica	PORTUGUESE Portugallica	RUMANIAN Rumenica	SERBIAN Servica	SLOVAK Slovaca	SWEDISH Suecica
polemika	polêmica	scriere polemică	polemičan spis (полемичан спис)	polemický spis polemika	dispyt ordstrid polemik
wielojęzyczny	poliglota	poliglot	mnogojezičan (многојезичан)	viacjazyčný viacrečový	polyglott
portret	retrato	portret	portret (портрет)	portrét	porträtt
pośmiertny	póstumo	postum	posmrtni (посмртни)	posmrtný	efterlämnad postum
dopisek postscriptum	pós-escrito	post-scriptum	dodatak (додатак) postskriptum (постскриптум)	postskriptum	efterskrift
przedmowa słowo wstępne wstęp	prefácio proémio	prefață	predgovor (предговор)	predslov	förord företal
przygotować	preparar	pregăti	pripraviti (приправити) pripremiti (припремити)	pripraviť prichystať	förbereda
przepis zasada reguła	prescrição	prescripție regulă regulament instrucție	propis (пропис) pravilo (правило)	predpis pravidlo	föreskrift reglement regel
przewodniczący prezes	presidente	preşedinte	predsednik (председник)	predseda prezident	president
prasa	imprensa prensa	presă tipar	štampa (штампа) tisak (тисак)	tlač	press

125

	ENGLISH Anglica	FRENCH Gallica	GERMAN Germanica	RUSSIAN Russica	SPANISH Hispanica
261	print[1]	imprimer tirer	drucken	pečatať (печатать) ottiskivať (оттискивать)	imprimir estampar
	print[2] →printed matter impression				
262	printed matter (print[2])	imprimé	Drucksache	pečatnoe (печатное)	impreso impresión

263	printed music	musique imprimée	Musikalien *pl.*
			Noten *pl.*
				noty (ноты) *pl.*
				muzykaľnoe
				  izdanie
				(музыкальное
				  издание)
					partitura

264	printer	imprimeur	Buchdrucker Drucker	tipograf (типограф) pečatnik (печатник)	impresor
265	printer's error (misprint typographical error)	erreur/faute typo- graphique	Druckfehler	opečatka (опечатка)	error de imprenta error tipográfico
	printing→ impression				
266	printing office (press)	imprimerie	Druckerei Druckerwerkstatt	tipografija (типография)	imprenta tipografia
	proceedings →transactions				
	proof→correction				
267	prose	prose	Prosa	proza (проза)	prosa
268	provide (with) (supply)	pourvoir munir	versehen	snabžať (снабжать)	proveer
269	pseudonym (pen-name)	pseudonyme nom de plume	Pseudonym	psevdonim (псевдоним)	seudónimo
	publication→edition				
	publication discontinued→ ceased publication				

BULGARICAN Bulgarica	CROATIAN Croatica	CZECH Bohemica	DANISH Danica	DUTCH Hollandica	FINNISH Fennica
pečatam (печатам) otpečatvam (отпечатвам)	tiskati štampati	tisknout	trykke	drukken	painaa
pečatno (proizvedenie) (печатно произведение)	tiskanica otisak	tisk otisk	tryksag præg	drukwerk	painotuote
notni izdanija (нотни издания) *pl.*	note tiskane muzikalije *pl.*	hudebniny *pl.*	musikalier *pl.* noder *pl.*	muziekboeken *pl.*	nuotit sävelmäjulkaisu
pečatar (печатар)	tiskar štampar	knihtiskař	bogtrykker	trykfeil drukker	kirjanpainaja painaja
pečatna greška (печатна грешка)	tiskarska pogreška	tisková chyba	trykfejl	drukfout	painovirhe
pečatnica (печатница)	tipografija štamparija	tiskárna	trykkeri bogtrykkeri	drukkerij	paino kirjapaino
proza (проза)	proza	próza	prosa	proza	proosa
snabdjavam (снабдявам)	snabdjeti	vybavit opatřit	forsyne	voorzien verzorgen	varustaa
psevdonim (псевдоним)	pseudonim	pseudonym	pseudonym forfatternavn	pseudoniem	salanimi pseudonyymi

127

	GREEK Graeca	HUNGARIAN Hungarica	ITALIAN Italica	LATIN Latina	NORWEGIAN Norvegica
261	ektypono *(ἐκτυπώνω)* typono *(τυπώνω)*	nyom (ki)nyomtat	imprimere stampare tirare	imprimere	trykke
262	entypon *(ἔντυπον)*	nyomtatvány	stampato	exemplar typis exscriptum impressum	trykksaker
263	partitura *(παρτιτούρα)*	kotta (nyomtatott) hangjegy zenemű	musica a stampa	nota musica impressa	musikalier *pl.* noter *pl.*
264	typographos *(τυπογράφος)*	nyomdász	tipografo stampatore	typographus	boktrykker
265	typographikon lathos *(τυπογραφικόν λάθος)*	sajtóhiba	errore di stampa	erratum typographi- cum	trykkfeil
226	typographeion *(τυπογραφεῖον)* typographia *(τυπογραφία)*	nyomda könyvnyomda	tipografia stamperia	officina typographica typographia	boktrykkeri trykkeri
267	pezos logos *(πεζός λόγος)*	próza	prosa	prosa	prosa
268	ephodiazo *(ἐφοδιάζω)* parecho *(παρέχω)*	ellát	provvedere	instruere	forsyne med ....
269	pseudonymon *(ψευδώνυμον)*	álnév	pseudonimo	pseudonymus	(antaget) forfatternavn pseudonym

POLISH Polonica	PORTUGUESE Portugallica	RUMANIAN Rumenica	SERBIAN Servica	SLOVAK Slovaca	SWEDISH Suecica
drukować	estampar	imprima tipări	štampati (штампати) tiskati (тискати)	tlačiť	trycka
dru(cze)k	impressão	imprimat tipăritură	tiskanica (тисканица) otisak (отисак)	odtlačok tlačivo	tryckalster trycksak
nuty *pl.* wydawnictwo nutowe	partitura	note imprimate *pl.*	muzikalije (музикалије) *pl.*	hudobniny *pl.*	musikalier *pl.*
drukarz	impressor	tipograf	štampar (штампар)	tlačiar	tryckeri tryckare
błąd drukarski	falha tipográfica	greşeală de tipar	štamparska greška (штампарска грешка)	tlačová chyba	tryckfel
drukarnia	imprensa tipografia	tipografie imprimerie	štamparija (штампарија) tipografija (типографија)	tlačiareň	tryckeri
proza	prosa	proză	proza (проза)	próza	prosa
przejrzeć	prover	înzestra	snabdeti (снабдети)	vybaviť opatriť zásobiť	förse
pseudonim	pseudónimo	pseudonim	pseudonim (псеудоним)	pseudonym	pseudonym

	ENGLISH Anglica	FRENCH Gallica	GERMAN Germanica	RUSSIAN Russica	SPANISH Hispanica
270	public library	bibliothèque publique bibliothèque de lecture publique	Volksbibliothek	publičnaja/ massovaja biblioteka (публичная/ массовая библиотека)	biblioteca (de lectura) pública
271	publish	publier éditer	verlegen	izdavať (издавать)	publicar editar
272	*be* published (appear come out)	paraître	erscheinen	izdavaťsja (издаваться) vychodiť (выходить)	publicarse salir
273	publisher[1]	éditeur	Verleger	izdateľ	editor
	publisher[2] →publishing house				
274	publishing house (publisher[2])	maison d'édition	Verlagsanstalt Verlag	izdateľstvo (издательство)	casa editora editorial
275	rare (scarce)	rare	selten	redkij (редкий)	raro
	recent→new				
276	record[1] (gramophone record)	disque	Schallplatte	grammofonnaja plastinka (граммофонная пластинка)	disco
	record[2] →minutes				
	record office →archives				
277	recto (right-hand page)	recto belle page	Recto Vorderseite Schauseite	pravaja/nečëtnaja stranica (правая/нечётная страница)	recto
	reference book →reference work				

BULGARIAN Bulgarica	CROATIAN Croatica	CZECH Bohemica	DANISH Danica	DUTCH Hollandica	FINNISH Fennica
obštoobra- zovatelna/masova biblioteka (общообразовател- на/масова библиотека)	javna knjižnica	lidová knihovna	folkebibliotek	openbare bibliotheek	yleinen kirjasto
izdavam (издавам)	izdavati	vydávat	forlægge	uitgeven	kustantaa
izlizam (излизам) izdavam se (издавам се)	objavljivati izlaziti	vycházet vyjít	udkomme	verschijnen	ilmestyä
izdatel (издател)	izdavać	vydavatel nakladatel	forlægger	uitgever	kustantaja
knigoizdatelstvo (книгоиздателство)	nakladno poduzeće	nakladatelství vydavatelství	forlag	uitgeverij	kustannusliike
rjadâk (рядък)	rijedak	vzácný	sjælden	zeldzaam	harvinainen
gramofonna ploča (грамофонна плоча)	gramofonska ploča	gramofonová deska	grammofonplade	grammofoonplaat	äänilevy
lice na list (лице на лист)	prva strana	přední strana recto	forside	voorzijde recto eerste bladzijde	(lehden) etusivu

		GREEK Graeca	HUNGARIAN Hungarica	ITALIAN Italica	LATIN Latina	NORWEGIAN Norvegica
270		laïke biliotheke (λαϊκή βιβλιοθήκη)	közművelődési könyvtár népkönyvtár közkönyvtár	biblioteca pubblica	bibliotheca popularis	folkebibliotek
271		ekdido (ἐκδίδω) demosieuo (δημοσιεύω)	kiad	pubblicare	edere publicare	forlegge utgive
272		ekdidomai (ἐκδίδομαι) ektyponomai (ἐκτυπώνομαι)	megjelenik	uscire essere pubblicato	edi publicari	utkomme
273		ekdotes (ἐκδότης)	kiadó könyvkiadó	editore	editor redemptor	forlegger utgiver
274		ekdotikos oikos (ἐκδοτικός οἶκος)	kiadóvállalat könyvkiadó vállalat	casa editrice	aedes librariae editor	forlag
275		spanios (σπάνιος)	ritka	raro	rarus	sjelden
276		diskos (δίσκος)	hanglemez	disco	discus grammo- phonii	grammofonplate
277		pleura prote (πλευρά πρώτη)	rektó	recto	recto	forside

POLISH Polonica	PORTUGUESE Portugallica	RUMANIAN Rumenica	SERBIAN Servica	SLOVAK Slovaca	SWEDISH Suecica
publiczna biblioteka powszechna	biblioteca (de leitura) pública	bibliotecă populară	javna biblioteka (јавна библиотека)	ľudová knižnica	folkbibliotek
wydawać	publicar	edita	izdavati (издавати)	vydávať	förlägga
ukazać się	aparecer sair	ieşi de sub tipar publica, se	objavljivati (објављивати) izlaziti (излазити)	vychádzať vyjsť	utkomma
wydawca	editor	editor	izdavač (издавач)	vydavateľ nakladateľ	förläggare
wydawnictwo	casa editora	editură	izdavačko preduzeće (издавачко предузеђе)	vydavateľstvo	bokförlag förlag
rzadki	raro	rar	redak (редак)	vzácny	sällsynt
płyta	disco	disc	gramofonska ploča (грамофонска плоча)	gramofónová platňa	grammofonskiva
recto	anverso recto	faţă recto	prva strana (прва страна)	lícna/prvá strana	framsida högersida recto

	ENGLISH Anglica	FRENCH Gallica	GERMAN Germanica	RUSSIAN Russica	SPANISH Hispanica
278	reference card (cross-reference card)	fiche de renvoi	Verweisungszettel Verweiszettel	ssyločnaja kartočka (ссылочная карточка) ssylka (ссылка)	ficha de referencia
279	reference library	bibliothèque de consultation sur place bibliothèque de référence	Handbücherei Präsenzbibliothek Nachschlagebiblio- thek	spravočnaja biblioteka (справочная библиотека)	biblioteca de consul- ta biblioteca de refe- rencia
280	reference work (reference book)	ouvrage de référence	Nachschlagewerk	spravočnik (справочник)	obra de consulta
281	refutation	réfutation	Widerlegung	oproverženie (опровержение)	refutación
	regulation →statute				
	register →list, index				
	reissue →reprint[1]				
	remains *pl.* →literary remains				
	removal slip →temporary card				
282	repertory	répertoire	Repertorium	ukazateľ (указатель)	repertorio
283	report[1] (account)	rapport bulletin	Bericht	doklad (доклад) otčët (отчёт) soobščenie (сообщение)	relación relato informe
	report[2] →minutes				
284	represent (illustrate)	illustrer figurer	abbilden	illjustrirovať (иллюстрировать) izobražať (изображать)	ilustrar figurar

BULGARIAN Bulgarica	CROATIAN Croatica	CZECH Bohemica	DANISH Danica	DUTCH Hollandica	FINNISH Fennica
prepratka (препратка)	uputnica uputni listić	odkazový lístek	henvisningskort	verwijzingskaart	viitekortti viittauskortti
podrâčna/ spravočna sbirka (подръчна/ справочна сбирка)	priručna biblioteka	příruční knihovna	håndbibliotek	handbibliotheek	käsikirjasto
spravočna kniga (справочна книга) spravočnik (справочник)	priručno djelo priručna knjiga	příručkové dílo	referencearbejde	naslagwerk	hakuteos käsikirja
oproverženie (опровержение)	opovrgnuće demanti	vyvrácení dementi	gendrivelse	tegenspraak weerlegging	vastatodistus
repertoar (репертоар)	repertoar	repertorium rejstřík	repertoire	repertorium	repertorio
doklad (доклад) otčet (отчет)	izveštaj referat	referát zpráva	beretning rapport	bericht mededeling rapport	ilmoitus kertomus selonteko selostus
iljustriram (илюстрирам) izobrazjavam (изобразявам)	predstavljati prikazivati	vyobrazit znázornit	illustrere fremstille	voorstellen	esittää kuvata

		GREEK Graeca	HUNGARIAN Hungarica	ITALIAN Italica	LATIN Latina	NORWEGIAN Norvegica
278		deltion anaphoras/ parapemptikon (δεκτίον ἀναφορᾶς/ παραπεμπτικόν)	utalócédula	scheda di rinvio	charta/scida referentiae	henvisningskort
279		procheiros bibliotheke (πρόχειρος βιβλιοθήκη)	kézikönyvtár	consultazione	bibliotheca manualis	håndbib:iotek
280		boethema (βοήθημα)	tájékoztatási segédlet referenszkönyv referenszmű	opera di consultazione	manuale refentiae liber auxiliarius	referansearbeid referanseverk
281		diapseusis (διάψευσις)	cáfolat	confutazione	confutatio refutatio	dementi gjendrivelse
282		heureterion (εὑρετήριον)	repertórium	repertorio	repertorium	repertoar
283		anaphora (ἀναφορά) angelia (ἀγγελία	jelentés	rapporto relazione	indicium refer(a)tum	forhandlinger innberetning rapport referat
284		apeikonizo (ἀπεικονίζω) diagrapho (διαγράφω) perigrapho (περιγράφω)	ábrázol	illustrare raffigurare rappresentare	describere	avbilde forme framstille

136

POLISH Polonica	PORTUGUESE Portugallica	RUMANIAN Rumenica	SERBIAN Servica	SLOVAK Slovaca	SWEDISH Suecica

odsyłacz
karta odsyłaczowa
	ficha remissiva	fişă de trimitere	uputni kataloški
			  listić
			(упутни каталош-
			  ки листић)
				odkazový lístok	hänvisningskort

księgozbiór
podręczny
	biblioteca de
	  referência/
	  consulta
		bibliotecă de con-
		  sultaţie
			priručna
			  biblioteka
			(приручна
			  библиотека)
				príručná knižnica	referensbibliotek

wydawnictwo informacyjne	obra de referência	carte de referinţă	priručnik (приручник)	príručkové dielo	uppslagsverk
odparcie	refutação	refutare refutaţie	opovrgavanje (оповргавање) demanti (деманти)	vyvrátenie dementi	vederläggning
repertorium	repertorio	repertoriu	repertoar (репертоар)	repertoár	repertoar
referat sprawozdanie	boletim	dare de seamă raport	izveštaj (извештај) referat (реферат)	správa	berättelse meddelande rapport
ilustrować	figurar ilustrar	ilustra reprezenta înfăţişa	ilustrovati (илустровати)	zobraziť znázorniť	illustrera

	ENGLISH Anglica	FRENCH Gallica	GERMAN Germanica	RUSSIAN Russica	SPANISH Hispanica
285	reprint[1] (reissue)	réimpression reproduction	Nachdruck Neudruck Wiederabdruck	pereizdanie (переиздание)	reimpresión reproducción
	reprint[2] →offprint				
286	reproduction (duplication) resume →summarize	reproduction	Reproduktion	reprodukcija (репродукция) razmnoženie (размножение)	reproducción
287	review n.	chronique revue	Rundschau	obzor (обзор) obozrenie (обозрение)	revista
288	review v.	rendre compte (de) recenser	besprechen rezensieren	recenzirovať (рецензировать)	reseñar
289	revise[1]	reviser revoir	revidieren	peresmatrivať (пересматривать) proverjať (проверять)	revisar
	revise[2] →rewrite				
290	revision	révision	Revision	peresmotr (пересмотр) proverka (проверка)	revisión
291	rewrite (revise rework)	remanier refondre retravailler	umarbeiten überarbeiten	pererabatyvať (перерабатывать)	refundir
	right-hand page →recto				
292	running title (running headline/ head)	titre courant	lebender Kolumnentitel Seitenüberschrift laufender Titel	kolontitul (колонтитул) kolontitulnaja stroka (колонтитульная строка)	título actual
	saying →adage				

BULGARIAN Bulgarica	CROATIAN Croatica	CZECH Bohemica	DANISH Danica	DUTCH Hollandica	FINNISH Fennica
novo (stereotipno) izdanie (ново стереотипно издание) prepečatvane (препечатване) preizdavane (преиздаване)	pretisak	otisk přetisk	optryk	herdruk nadruk	jälkipainos uusi painos
reprodukcija (репродукция) razmnožavane (размножаване)	reprodukcija	reprodukce rozmnožování	reproduktion	reproductie vermenigvuldiging	jäljennys monistus
pregled (преглед)	pregled	revue	revu	overzicht revue	katsaus
razgleždam (разглеждам) recenziram (рецензирам)	recenzirati prikazati	recenzovat posuzovat	anmelde recensere	recenseren	arvostella selostaa
pregleždam (преглеждам) reviziram (ревизирам)	pregledati	revidovat přehlédnout	revidere gennemse	nazien herzien	tarkastaa
pregled (преглед) revizija (ревизия)	pregled	revize přehlédnutí	revision	herziening	tarkistus tarkastus
prerabotvam (преработвам)	preraditi	přepracovat upravit	bearbejde omarbejde	omwerken	muokata uusia
kolontitul (колонтитул)	tekući naslov	průběžný název	levende kolum-netitel	kolomtitel paginatitel	vaihtuva sivu(n)-otsikko elävä sivu(n)-otsikko

	GREEK Graeca	HUNGARIAN Hungarica	ITALIAN Italica	LATIN Latina	NORWEGIAN Norvegica
285	anatyposis (ἀνατύπωσις) metatyposis (μετατύπωσις)	utánnyomás	ristampa	reimpressum	avtrykk opptrykk
286	anatypoma (ἀνατύπωμα) anaplasis (ἀνάπλασις)	reprodukció sokszorosítás	riproduzione	multiplicatio reproductio	reproduksjon
287	anaskopesis (ἀνασκόπησις)	szemle	rivista rassegna letteraria	acta *pl.* pagella publica	revy rundskue
288	krino (κρίνω) deloo (δηλόω)	ismertet tárgyal	recensire	recensere exponere	anmelde
289	epitheoro (ἐπιθεωρῶ) elencho (ἐλέγχω)	átnéz	rivedere	revidere recognoscere	gjennomse revidere
290	anatheoresis (ἀναθεώρησις)	átnézés revízió	revisione	revisio recognitio	revisjon
291	diaskeuazo (διασκευάζω) metapoieo (μεταποιέω)	átdolgoz	rifondere rielaborare ritoccare	denuo conscribere retractare	bearbeide omarbeide omdanne
292	epititlos selidos (ἐπίτιτλος σελίδος)	élőfej	titolo corrente a capo pagina	titulus currens	levende kolumntittel løpetittel

POLISH Polonica	PORTUGUESE Portugallica	RUMANIAN Rumenica	SERBIAN Servica	SLOVAK Slovaca	SWEDISH Suecica
przedruk	reimpressão	reimprimare	preštampavanje (прештампавање)	dotlač dotlačok	nytryck(ning) omtryck(ning)
powielanie reprodukcja	reprodução	multiplicare reproducere	reprodukcija (репродукција)	reprodukcia rozmnoženie	avtryck reproduktion
przegląd	revista	revistă cronică	pregled (преглед)	obzor	revy översikt krönika
omawiać recenzować	resenhar	recenza	recenzirati (рецензирати) prikazati (приказати)	recenzovať posúdiť	recensera anmäla
przeglądać	corrigir	revedea revizui	pregledati (прегледати)	prezrieť	revidera
rewizja	revisão	revizie	pregled (преглед)	revízia	revision
przerobić	refundir retocar	reface prelucra	preraditi (прерадити)	prepracovať prerobiť	omarbeta
żywa pagina	título corrente	colontitlu	tekući naslov (текући наслов)	priebežný názov	kolumntitel

	ENGLISH Anglica	FRENCH Gallica	GERMAN Germanica	RUSSIAN Russica	SPANISH Hispanica
293	scale	échelle	Maßstab	masštab (масштаб)	escala
	scarce →rare				
294	scenario (text for a film)	scénario	Drehbuch Textbuch eines Films	scenarij (сценарий) text filma (текст фильма)	guión
295	school	école	Schule	škola (школа)	escuela
296	science	sciences *pl.*	Wissenschaft	nauka (наука)	ciencia
297	science fiction	science-fiction	wissenschaftlich- -phantastischer Roman wissenschaftliche Abenteuergeschich- ten	naučno- -fantastičeskaja literatura (научно- -фантастическая литература) naučnaja fantastika (научная фантастика)	ciencia ficción
298	screen	trame	Raster	rastr (растр)	pantalla trama
299	script (writing)	écriture lettres *pl.*	Schrift	pis'mo (письмо) šrift (шрифт)	escritura letra
	secondary entry →added entry				
300	second-hand bookshop	librairie d'occasion	Antiquariat Antiquariatsbuch- handlung	bukinističeckij magazin (букинистический магазин)	librería de viejo y de ocasión
301	section	section	Sektion Abteilung	sekcija (секция) otdel (отдел) otdelenie (отделение)	sección parte

BULGARIAN Bulgarica	CROATIAN Croatica	CZECH Bohemica	DANISH Danica	DUTCH Hollandica	FINNISH Fennica
maštab (мащаб)	mjerilo	měřítko	målestok	schaal	mittakaava
scenarij (сценарий)	scenarij knjiga snimanja	scénář	drejebog	draaiboek	elokuvan käsikir- joitus elokuvakäsikir- joitus skenaario
učilište (училище) škola (школа)	škola	škola	skole	school	koulu
nauka (наука)	nauka	věda	videnskab	wetenschap	tiede
naučno- -fantastična literatura (научно- -фантастична литература) naučna fantastika (науча фан- тастика)	znanstveno-fanta- stična književnost	vědecko-fantas- tická literatura	science fiction	(wetenschappelij- ke) tokomstroman	tieteisromaani
raster (растер)	raster	rastr	raster	raster	rasteri
pisane (писане) šrift (шрифт)	pisanje	písmo psaní	skrift skrivemåde skrivning	schrift	kirjasin kirjoitus
antikvarna knižarnica (антикварна книжарница)	antikvarijat antikvarnica	antikvariát	antikvariat	tweedehandsboek- winkel	antikvariaatti
sekcija (секция) otdel(enie) (отделение)	sekcija odjeljak dio	sekce oddíl	del sektion	sectie deel gedeelte afdeling	jaosto sektio osasto pykälä

	GREEK Graeca	HUNGARIAN Hungarica	ITALIAN Italica	LATIN Latina	NORWEGIAN Norvegica
293	klimax *(κλῖμαξ)*	lépték mérték méretarány	scala	scala	målestokk
294	senarion *(σενάριον)*	forgatókönyv	scenario	scenarium	dreiebok
295	schole *(σχολή)*	iskola	scuola	schola	skole
296	episteme *(ἐπιστήμη)*	tudomány	scienza	scientia doctrina disciplina litterae *pl.*	vitenskap
297	epistemonikon phantastikon mythistorema *(ἐπιστημονικόν φανταστικόν μυθιστόρημα)* mythistorema epistemonikes phantasias *(μυθιστόρημα ἐπιστημονικῆς φαντασίας)*	tudományos fan- tasztikus irodalom sci-fi	romanzo di fantascienza	prosa scientifica phantastica	fremtidsromaner *pl.* science fiction
298	raster *(ράστερ)*	raszter	reticolo	raster	raster
299	graphe *(γραφή)* syngraphe *(συγγραφή)*	írás	scritto scrittura	scriptum scriptura	skrift skrivning
300	palaiobibliopoleion *(παλαιοβιβλιοπωλεῖον)*	antikvárium	libreria d'antiquariato	antiquarium	antikvariat
301	tmema *(τμῆμα)* meros *(μέρος)*	osztály rész szakasz	parte sezione	pars sectio	avsnitt

POLISH Polonica	PORTUGUESE Portugallica	RUMANIAN Rumenica	SERBIAN Servica	SLOVAK Slovaca	SWEDISH Suecica
skala	escala	scară	razmera (размера)	mierka	skala
scenariusz	guião	scenariu	scenario (сценарно)	scenár	scenario
szkoła	escola	şcoală	škola (школа)	škola	skola
nauka wiedza	ciência	ştiinţă	nauka (наука)	veda	vetenskap
literatura fantastyczno- naukowa	romance de anticipação científica	roman ştiinţific fantastic	naučno- -fantastični roman (научно- -фантастични роман)	fantastický roman vedecko-fantastic- ká litaratúra	science fiction
raster	tela	raster	raster (растер)	raster	raster
pismo	escrita	scriere scriptură scripte *pl.*	pisanje (писање) rukopis (рукопис)	písmo písanie	skrift
antykwariat	antiquário alfarrabista	anticariat	antikvarnica (антикварница)	antikvariát	antikvariat
sekcja oddział rozdział	secção parte	secţie parte	sekcija (секција) odeljak (одељак) deo (део)	sekcia oddiel odsek	sektion del avdelning

		ENGLISH Angliea	FRENCH Gallica	GERMAN Germanica	RUSSIAN Russica	SPANISH Hispanica
302		select (choose)	choisir sélectionner	wählen	vybirať (выбирать) otbirať (отбирать)	elegir escoger
303		selection (chrestomathy)	morceaux choisis chrestomathie	Auswahl ausgewählte Stücke/Werke	izbrannoe (избранное) chrestomatija (хрестоматия)	selección crestomatía
		separate →offprint				
304		sequel	suite	Fortsetzung	prodolženie (продолжение)	continuación
305		series	série suite	Serie Reihe	serija (серия)	serie
		session →sitting				
306		sew (stitch)	brocher coudre	heften	brošjurovať (брошюровать)	encuadernar en rústica
307		sheet	feuille	Bogen	list (лист)	hoja
		shelf mark/number →location mark				
308		shorthand (stenography)	sténographie	Kurzschrift Stenographie	stenografija (стенография) skoropis' (скоропись)	estenografía
309		short story (novelette tale)	nouvelle conte	Novelle Erzählung	novella (новелла) rasskaz (рассказ)	novela corta cuento
310		signature	signature	Signatur	signatura (сигнатура)	signatura

BULGARIAN Bulgarica	CROATIAN Croatica	CZECH Bohemica	DANISH Danica	DUTCH Hollandica	FINNISH Fennica
izbiram (избирам)	izabrati odabrati	vybírat	udvælge vælge	kiezen uitkiezen	valikoida valita

hristomatija
(христоматия)
izbrani otkâsi
(избрани откъси) *pl.*
sbornik ot
  izbrani
  proizvedenija
(сборник от
  избрани
  произведения)
	hrestomatija	chrestomatie
		výbor
			krestomati	chrestomathie	kirjallisuuden
					  näytteitä
					krestomatia

prodâlženie (продължение)	nastavak	pokračování	fortsættelse	vervolg	jatko
serija (серия) poredica (поредица)	serija	řada serie	serie række	reeks serie	jakso sarja
podšivam (подшивам)	broširati	brožovat sešívat	hæfte	innaaien	kiinnittää nitoa
kola (кола) list (лист)	arak	arch	ark	folio vel	arkki
stenografija (стенография)	stenografija	těsnopis	stenografi	kortschrift stenografie	pikakirjoitus
novela (новела) razkaz (разказ)	novela priča pripovijetka	novela	novelle	novelle	novelli kertomus
signatura (сигнатура)	signatura	signatura	signatur	signatuur	luokanmerkki paikanmerkki

	GREEK Graeca	**HUNGARIAN** Hungarica	**ITALIAN** Italica	**LATIN** Latina	**NORWEGIAN** Norvegica
302	eklego (ἐκλέγω)	válogat	assortire scegliere	deligere eligere seligere	velge ut
303	analekta (ἀνάλέκτα) pl. chrestomatheia (χρηστομάθεια)	szemelvénygyűjte- mény válogatás	crestomazia	chrestomathia electi loci pl. eclogarii pl.	krestomati
304	synecheia (συνέχεια)	folytatás	continuazione	continuatio	fortsettelse
305	seira (σειρά)	sorozat	serie	ordo series	følge rekke serie
306	syndeo (συνδέω)	fűz	legare (alla rustica)	(filo) compingere	hefte
307	kolla (κόλλα)	ív	foglio	arcus plagula	ark blad
308	stenographia (στενογραφία)	gyorsírás	stenografia	stenographia tachygraphia	stenografi
309	diegesis (διήγησις)	novella elbeszélés	novella	novella narratio	fortelling novelle
310	yposemeiosis (ὑποσημείωσις)	jelzet	segnatura	signatura	signatur

148

POLISH Polonica	PORTUGUESE Portugallica	RUMANIAN Rumenica	SERBIAN Servica	SLOVAK Slovaca	SWEDISH Suecica
wybierać	escolher seleccionar	alege	izabrati (изабрати) odabrati (одабрати)	vyberať preberať	utvälja
chrestomatia wypisy	escolha selecção	analecte *pl.* crestomatie	hrestomatija (хрестоматија)	chrestomatia výbor	krestomati urval
dalszy ciąg kontynuacja	continuação	continuare	nastavak (наставак)	pakračovanie	följd
seria	série	serie	serija (серија)	rad séria	serie
broszurować	brochar coser	broşa	broširati (броширати)	brožovať zošívať	häfta
arkusz	folha	coală foaie	arak (арак) tabak (табак)	hárok	ark
stenografia	estenografia	stenografie	stenografija (стенографија)	stenografia	stenografi
nowela	novela	nuvelă poveste	novela (новела) priča (прича)	novela	novell
sygnatura	assinatura	signatură	signatura (сигнатура)	signatúra	signatur

	ENGLISH Anglica	FRENCH Gallica	GERMAN Germanica	RUSSIAN Russica	SPANISH Hispanica
311	sitting (session)	séance assemblée	Sitzung	zasedanie (заседание) sessija (сессия)	sesión
312	sketch	esquisse croquis	Skizze Entwurf	ėskiz (эскиз) nabrosok (набросок) očerk (очерк)	esbozo diseño bosquejo
	small →little				
313	society	société	Gesellschaft	obščestvo (общество) associacija (ассоциация)	sociedad
	sole →only				
314	song	chant	Gesang	pesnja (песня)	canto
315	songbook (hymn-book hymnal)	chansonnier recueil de chansons livre de chant	Liederbuch Liederkranz Gesangbuch	pesennik (песенник) sbornik pesen (сборник песен)	cancionero
316	source	source	Quelle	istočnik (источник)	fuente
317	special bibliography (subject bibliography)	bibliographie spéciale/spécialisée	Fachbibliographie	predmetnaja/ specialnaja bibliografija (предметная/ специальная библиография)	bibliografía especializada
318	special library	bibliothèque spécialisée	Fachbibliothek	specialnaja biblioteka (специальная библиотека)	biblioteca especial(izada)
319	speech (oration)	discours allocution	Rede	reč' (речь)	alocución discurso oración
	spine →back				
320	stack-room (stacks pl. bookstacks pl.)	dépôt/magasin de livres	Büchermagazin Bücherspeicher Magazin	knigochranilišče (книгохранилище)	depósito de libros

BULGARIAN Bulgarica	CROATIAN Croatica	CZECH Bohemica	DANISH Danica	DUTCH Hollandica	FINNISH Fennica
zasedanie (заседание) sesija (сесия)	sednica zasedanje	zasedání	session møde	zitting	istunto
skica (скица) očerk (очерк) skeč (скеч)	skica	náčrtek nákres nástin	skitse udkast	schets	alustelma luonnos suunnitelma
družestvo (дружество)	društvo	společnost	selskab samfund	geselschap maatschappij	seura
pesen (песен)	pjesma	píseň	sang	zang	laulu
pesnopojka (песнопойка) sbornik pesni (сборник песни)	pjesmarica	zpěvník kancionál	sangbog	liederboek	laulukirja
izvor (извор) iztočnik (източник)	izvor	pramen zdroj	kilde ophav	bron	lähde

otraslova/
  specialna
  bibliografija
(отраслова/
  специална
  библиография)
	stručna bibliogra-
	fija
		speciální biblio-
		grafie
			fagbibliografi	speciale
				  bibliografie
				vakbibliografie
					erikoisbibliografia

otraslova/
  specialna
  biblioteka
(отраслова/
  специална
  библиотека)
	stručna knjižnica	speciální knihovna	fagbibliotek	speciale bibliotheek
				vakbibliotheek
					erikoiskirjasto

reč (реч) slovo (слово)	govor riječ	řeč	tale	rede	puhe
knigohranilište (книгохранилище)	skladište knjiga	skladiště knih	bogmagasin	boekenmagazijn	kirjavarasto

	GREEK Graeca	HUNGARIAN Hungarica	ITALIAN Italica	LATIN Latina	NORWEGIAN Norvegica
311	synedriasis *(συνεδρίασις)*	ülés(szak)	sessione	consessus sessio	møte sesjon
312	skitso *(σκίτσο)* schedion *(σχέδιον)*	vázlat	schizzo	adumbratio descriptio planum	grunnriss plan skisse
313	hetairia *(ἑταιρία)*	társulat társaság	società	societas	selskap samfunn
314	tragudi *(τραγούδι)* melos *(μέλος)* ode *(ᾠδή)*	ének	canto	cantus carmen	sang vers vise
315	sylloge tragudion *(συλλογή τραγουδιῶν)*	daloskönyv énekeskönyv	canzoniere	liber canticorum/ carminum cantionale	sangbok
316	pege *(πηγή)*	forrás	fonte	fons	kilde opphav oppkomme
317	eidike bibliographia *(εἰδική βιβλιογραφία)*	szakbibliográfia	bibliografia speciale	bibliographia specialis	fagbibliografi
318	eidike bibliotheke *(εἰδική βιβλιοθήκη)*	szakkönyvtár	biblioteca speciale	bibliotheca specialis	fagbibliotek
319	logos *(λόγος)* homilia *(ὁμιλία)*	beszéd szónoklat	discorso orazione	dictio sermo oratio	språk tale
320	apotheke biblion *(ἀποθήκη βιβλίων)*	könyvraktár	magazzino librario	horreum librorum	bokmagasin

POLISH Polonica	PORTUGUESE Portugallica	RUMANIAN Rumenica	SERBIAN Servica	SLOVAK Slovaca	SWEDISH Suecica
posiedzenie sesja	sessão	şedinţă sesiune	sednica (седница) zasedanje (заседање)	zasadanie zasadnutie	sammanträde
szkic zarys plan	debuxo esboço	schiţă	skica (скица)	náčrt nárys nákres	skiss plan
towarzystwo	sociedade	societate	društvo (друштво)	spoločnosť	sällskap
śpiew pieśń	cantiga canção	cîntec cînt	pesma (песма)	pieseň	sång
śpiewnik zbiór pieśni kancjonal	cancioneiro	carte/culegere de cîntece	pesmarica (песмарица)	spevník kancionál	sångbok visbok
żródło	fonte	izvor sursă	izvor (извор)	prameň	källa
bibliografia specjalna	bibliografia especializada	bibliografie specială	stručna bibliografija (стручна библогриафија)	odborná bibliogra- fia	ämnesbibliografi fackbibliografi
biblioteka specjalna	biblioteca especializada	bibliotecă specială	specijalna biblioteka (специјална библиотека)	odborná knižnica	specialbibliotek
mowa pzemówenie	dicção discurso oração	cuvîntare oraţiune	govor (говор) reč (реч)	reč	tal
magazyn biblioteczny	depósito de livros	magazie de cărţi	magazin za knjige (магазин за књиге)	sklad kníh	bokmagasin

153

	ENGLISH Anglica	FRENCH Gallica	GERMAN Germanica	RUSSIAN Russica	SPANISH Hispanica
	stacks →**stack-room**				
321	**standardisation**	normalisation	Normung Standardisierung	normalizacija (нормализация) standartizacija (стандартизация)	normalización estandartización
	stanza →**strophe**				
322	**state library**	bibliothèque d'État	Staatsbibliothek	gosudarstvennaja biblioteka (государственная библиотека)	biblioteca de Estado
323	**statute**[1] **(regulation)**	statuts *pl.* réglement	Statut	pravila (правила) *pl.* ustav (устав) statut (статут)	estatuto
	statute[2] →**law** **stenography** →**shorthand** **stitch** →**sew**				
324	**story** **(tale)**	récit conte	Erzählung	rasskaz (рассказ) novella (новелла)	novela corta cuento narración relato
	strip cartoon →**comic strip**				
325	**strophe** **(stanza)**	strophe	Strophe	strofa (строфа)	estrofa
326	**study** **(paper**[2]**)**	étude	Studie	issledovanije (исследование) statja (статья) ètjud (этюд)	estudio
	subject bibliography →**special** **bibliography**				
327	**subject heading**	mot-rubrique mot-centre	Schlagwort	predmetnyj zagolovok (предметный заголовок)	palabra de título
328	**subscription**	abonnement	Abonnement laufender Bezug	podpiska (подписка)	servicio de abono subscripción

BULGARIAN Bulgarica	CROATIAN Croatica	CZECH Bohemica	DANISH Danica	DUTCH Hollandica	FINNISH Fennica
normalizacija (нормализация) standartizacija (стандартизация)	normiranje standardizacija	normalizace	standardisering	normalisatie standardisatie	normaalistaminen

dâržavna
  biblioteka
(държавна
  библиотека)
	državna knjižnica	státní knihovna	statsbibliotek	staatsbibliotheek	valtionkirjasto

pravilnik (правилник) ustav (устав)	pravilo propis statut	pravidlo předpis statut	statut regel	regel voorschrift statuut	ohje sääntö
razkaz (разказ)	priča pripovijetka pripovijest	povídka	fortælling novelle	verhaal vertelling	kertomus kertoelma
strofa (строфа) kuplet (куплет)	strofa kitica	sloka strofa	strofe	strofe	säkeistö stroofi
studija (студия) proučvane (проучване) etjud (етюд)	studija ogled	studie	essay	studie artikel	tutkielma

zaglavna duma
(заглавна дума)
predmetna
  rubrika
(предметна
  рубрика)
	predmetna oznaka	předmětové heslo	slagord	onderwerpswoord
				trefwoord
					hakusana

abonament (абонамент) podpiska (подписка)	pretplata	předplatné	abonnement	abonnement	ennakkotilaus

155

	GREEK Graeca	HUNGARIAN Hungarica	ITALIAN Italica	LATIN Latina	NORWEGIAN Norvegica
321	typopoiesis (τυποποίησις)	szabványosítás	normalizzazione standardizzazione	normalisatio	standardisering
322	kratike bibliotheke (κρατική βιβλιοθήκη)	állami könyvtár	biblioteca statale	bibliotheca publica	statsbibliotek
323	kanon (κανών) katastatikon (καταστατικόν)	alapszabály szabály(zL t)	precetto regola(mento) statuto	regula praescriptum praeceptum	forskrift regel
324	diegema (διήγημα)	elbeszélés	narrazione racconto novella	fabula narratio novella	fortelling
325	strophe (στροφή)	versszak	strofa	stropha	strofe vers
326	melete (μελέτη)	tanulmány dolgozat	studio	studium	avhandling essay studium
327	lemma (λῆμμα)	címszó	parola d'ordine per soggetto	vox vocabulum	slagord
328	syndrome (συνδρομή)	előfizetés	abbonamento	praenumeratio	abonnement subskripsjon

156

POLISH Polonica	PORTUGUESE Portugallica	RUMANIAN Rumenica	SERBIAN Servica	SLOVAK Slovaca	SWEDISH Suecica
normalizacja	normalização	normalizaţie	normiranje (нормирање) standardizacija (стандардизација)	normovanie	standardisering
biblioteka państwowa	biblioteca do Estado	bibliotecă de stat	državna biblioteka (државна библиотека)	štátna knižnica	statligt bibliotek
ustawa statut przepis reguła	estatuto norma regra	statut regulă regulament	pravilo (правило) propis (пропис) statut (статут)	pravidlo predpis štatút	statut regel
opowiadanie nowela	novela	povestire nuvelă	pripovetka (приповетка) priča (прича)	poviedka novela	beträttelse novell
strofa	estrofa	strofă	strofa (строфа) kitica (китица)	sloha strofa	strof
studia studium	ensaio estudo	studiu	studija (студија) ogled (оглед)	štúdia	studie
temat	encabeçamento palavra de título	cuvînt-titlu	predmetna odrednica (предметна одредница)	pojem	slagord
prenumerata	subscripçã o	abonament	pretplata (претплата)	predplatné	abonnemang

157

	ENGLISH Anglica	FRENCH Gallica	GERMAN Germanica	RUSSIAN Russica	SPANISH Hispanica
329	summarize (sum up resume)	résumer	zusammenfassen	podvodiť itogi (подводить итоги) rezjumirovať (резюмировать)	recoger reunir resumir
330	summary (abstract) sum up →summarize	résumé sommaire	Zusammenfassung Resumé	rezjume (резюме)	resumen
331	supplement (appendix) supplement v. →complete v.	supplément annexe	Beiblatt Beilage	priloženie (приложение) dopolnenie (дополнение)	adjunto suplemento anejo
332	supplementary supply →provide	supplémentaire	Ergänzungs-	dopolnitel'nyj (дополнительный)	suplementario
333	table	tableau	Tabelle	tablica (таблица)	tabla
334	table of contents tale →fable (short) story	sommaire table des matières	Inhaltsverzeichnis	oglavlenie (оглавление)	índice tabla de materias sumario
335	on tape	sur bande	auf Tonband auf Magnetband auf Magnetophon	na magnitofonnoj lente (на магнитофонной ленте)	en cinta
336	temporary card (temporary/removal slip)	fiche provisoire	Interimsaufnahme provisorischer Zettel Interimszettel	vremennaja kartočka (временная карточка)	ficha provisional
337	text (words) pl	texte paroles pl.	Text	tekst (текст)	texto
338	textbook (manual)	livre d'étude livre scolaire	Lehrbuch	učebnik (учебник) učebnaja kniga (учебная книга)	libro de texto manual (escolar)

BULGARIAN Bulgarica	CROATIAN Croatica	CZECH Bohemica	DANISH Danica	DUTCH Hollandica	FINNISH Fennica
rezjumiram (резюмирам)	rezimirati ukratko izložiti	shrnout resumovat	sammenfatte	samenvatten	koota yhdistää
rezjume (резюме)	rezime	resumé shrnutí	sammenfatning	samenvatting	yhteenveto
priloženie (приложение) priturka (притурка)	dodatak prilog	dodatek příloha	bilag tillæg	bijlage bijvoegsel	liite lisälehti
dopâlnitelen (допълнителен)	dopunski	dodatkový	tillægs-	supplement-	lisä-
tablica (таблица)	tabela	tabulka	tabel	tabel	taulukko
sâdâržanie (съдържание)	sadržaj kazalo	obsah rejstřík ukazatel	indholdsforteg- nelse	inhoudsopgave	sisällysluettelo
na lenta (на лента)	na magnetofon- skoj vrpci	na pásce	på bånd	op toonband op de band	nauhalle nauhaan
vremenna kartička (временна картичка)	privremeni listić	předběžný/proza- tímní lístek	foreløbig seddel	voorlopige kaart	väliaikaiskortti tilapäiskortti tilapäislippu
tekst (текст) dumi (думи) pl.	tekst	text	tekst	tekst	teksti
učebnik (учебник)	udžbenik	učebnice	skolebog lærebog	leerboek schoolboek	koulukirja oppikirja

159

	GREEK Graeca	HUNGARIAN Hungarica	ITALIAN Italica	LATIN Latina	NORWEGIAN Norvegica
329	synopsizo *(συνοψίζω)*	összefoglal	riassumere compendiare	complecti comprehendere	gjenoppta sammenfatte
330	perilepsis *(περίληψις)*	összefoglalás	sommario	complexio	sammendrag
331	parartema *(παράρτημα)*	melléklet	annesso	addenda additamentum advolutum	bilag tillegg
332	sympleromatikos *(συμπληρωματικός)*	pót- kiegészítő	suppletivo suppletorio	suppletivus supplementum	tillegs-
333	pinax *(πίναξ)*	táblázat	tabella	tabella	tabell tavle
334	pinax periechomenon *(πίναξ περιεχομένων)*	tartalomjegyzék	sommario indice delle materie	index elenchus argumentum	innhold innholdsfortegnelse register
355	sten phonoleptike tainia *(στήν φωνοληπτική* *ταινία)* sten phonotainian *(στήν φωνοταινίαν)*	hangszalagon szalagon	su nastro	taenia pellicularis vocis	på bånd/lydbånd lydopptak på bånd
336	prosorinon deltion *(προσωρινόν δελτίον)* prosorine katalogogra- phesis *(προσωρινή* *καταλογογράφησις)*	ideiglenes címleírás	scheda provvisoria	charta/scida tempora- ria	foreløpig seddel/kort
337	keimenon *(κείμενον)*	szöveg	testo	contextus textus	tekst
338	didaktikon biblion *(διδακτικόν βιβλίον)* encheiridion *(ἐγχειρίδιον)*	tankönyv	libro scolastico	liber scholasticus rudimentum	lærebok skolebok

POLISH Polonica	PORTUGUESE Portugallica	RUMANIAN Rumenica	SERBIAN Servica	SLOVAK Slovaca	SWEDISH Suecica
streszczać	recolher resumir	rezuma	rezimirati (резимирати) ukratko izložiti (укратко изло- жити)	zhrnúť rezumovať	sammanfatta
streszczenie	resumo	rezumat	rezime (резиме)	resumé zhrnutie	resumé sammanfattning
dodatek załącznik	anexo suplemento	anexă supliment	dodatak (додатак) prilog (прилог)	príloha	bilaga bihang tillägg
dodatkowy dopełniający	suplementário	suplimentar	dopunski (допунски)	dodatkový	tillägs- supplement-
tabela	tabela	tabel tablă	tabela (табела)	tabuľka tabela	tabell
spis rzeczy/treści	índice	tablă de materii sumar	sadržaj (садржај) kazalo (казало)	zoznam obsah	innehållsförteck- ning
na taśmie	sobre banda magnética	pe bandă de magnetofon	na magnetofon- skoj traci (на магнетофон- ској траци)	na páske	på band
karta zastępcza	ficha provisória	fişă provizorie	privremeni kata- loški listiç (привремени ката- лошки листић)	dočasný lístok	provisoriskt kort
tekst	texto	text	tekst (текст)	text	text
podręcznik szkolny	livro escolar	carte de şcoală manual	udžbenik (уџбеник)	učebnica	lärobok

		ENGLISH Anglica	FRENCH Gallica	GERMAN Germanica	RUSSIAN Russica	SPANISH Hispanica
339		the	l' le la *pl.* les	der die das *pl.* die	—	el la lo *pl.* los las
340		thesis (dissertation)	thèse dissertation	Dissertation	dissertacija (диссертация)	disertación
341		title	titre intitulé	Titel	titul (титул) zaglavie (заглавие) zagolovok (заголовок)	título
342		title entry	notice sous le titre	Titeleintragung Eintragung unter dem Sachtitel	opisanie pod zaglaviem (описание под заглавием)	asiento de título
343		title page	page/feuille de titre	Titelseite Titelblatt	titulnyj/ zaglavnyj list (титульный/ заглавный лист)	portada
344		to	á	zu an bis	k, ko (к, ко)	a
		tome→volume				
345		tragedy	tragédie	Tragödie Trauerspiel	tragedija (трагедия)	tragedia
346		transactions *pl.* (proceedings *pl.* bulletin)	actes *pl.* bulletin (d'informa- tions)	Mitteilungen *pl.*	izvestija (известия) *pl.* (naučnye) zapiski (научные записки) *pl.* trudy (труды) *pl.*	comunicado comunicación
		transcript →copy[1]				
347		transcription	transcription	Transkription	transkripcija (транскрипция)	transcripción
348		translate	traduire	übersetzen	perevodiť (переводить)	traducir

162

BULGARIAN Bulgarica	CROATIAN Croatica	CZECH Bohemica	DANISH Danica	DUTCH Hollandica	FINNISH Fennica
-ât (-ът), -a; -jat (-ят), -ja (-я); *pl.* -te (-те) -ta (-та); *pl.* -te (-те) -to (-то); *pl.* -ta (-та)	—	—	den, det; *pl.* de	de het	—
disertacija (дисертация) teza (теза)	disertacija	disertace	disputats afhandling	dissertatie proefschrift	väitöskirja dissertaatio
zaglavie (заглавие)	naslov titula natpis	nadpis titul název	titel	titel	otsikko nimi titteli
opisanie pod zaglavie (описание под заглавие)	stvarni naslov	popis pod názvem	sagtitel	titelkaart opneming onder de titel	nimekkeenmukai- nen kirjaus
zaglavna stranica (заглавна страница)	naslovni list	titulní list	titelblad	titelblad titelpagina	nimiölehti
do (до) na (на)	do k ka	k ke ku	til	te tot	luo(kse) kohti -lle
tragedija (трагедия)	tragedija	tragédie truchlohra	sørgespil tragedie	treurspel tragedie	murhenäytelmä tragedia
izvestija (известия) *pl.* trudove (трудове) *pl.*	izvještaj saopćenja *pl.* bilten	zprávy *pl.* sdělení	meddelelse meldinger *pl.* skrifter *pl.*	mededeling	ilmoitus tiedonanto tiedotus
transkripcija (транскрипция)	transkripcija	transkripce	transskription	transcriptie	siirtokirjoitus
prevеždam (превеждам)	prevoditi tumačiti	překládat	oversætte	vertalen	kääntää

	GREEK Graeca	HUNGARIAN Hungarica	ITALIAN Italica	LATIN Latina	NORWEGIAN Norvegica
339	ho *(ὁ) pl.* hoi *(οἱ)* he *(ἡ)* hai *(αἱ)* to *(τό)* ta *(τά)*	a, az	il *pl.* i l' — lo gli la le	—	den det *pl.* de
340	didaktorike diatribe *(διδακτορική* *διατριβή)*	értekezés disszertáció	dissertazione	dissertatio	disputas
341	titlos *(τίτλος)*	cím	titolo	titulus inscriptio libri	tittel
342	ypotitlos *(ὑποτίτλος)*	tárgyi címszó	parola d'ordine per titoli	vocabulum rei/ob- iecti/tituli	emneord saktitel
343	prometopis *(προμετωπίς)*	címlap	frontespizio	pagina tituli	tittelblad
344	hos *(ὡς)* pros *(πρός)* epi *(ἐπί)*	-hoz, -hez, -höz -nak, -nek	a ad	ad	til
345	tragodia *(τραγῳδία)*	tragédia szomorújáték	tragedia	tragœdia	sørgespill tragedie
346	anakoinosis *(ἀνακοίνωσις)* metadosis *(μετάδοσις)* deltion *(δελτίον)*	közlemények *pl.* akták *pl.* értesítő	comunicazione pubblicazione	communicationes *pl.* acta *pl.*	meddelelse kommuniké skrifter *pl.*
347	metagraphe *(μεταγραφή)*	átírás	trascrizione	transcriptio	transskripsjon
348	metaphrazo *(μεταφράζω)*	fordít	tradurre interpretare	convertere transferre vertere interpretare	oversette

164

POLISH Polonica	PORTUGUESE Portugallica	RUMANIAN Rumenica	SERBIAN Servica	SLOVAK Slovaca	SWEDISH Suecica
—	o *pl.* os a as	-(u)l *pl.* -i -a -le -le	—	—	den det *pl.* de
dysertacja	dissertação	disertaţie teză	disertacija (дисертација)	dizertácia	dissertation akademisk avhandling
tytuł	título	titlu	naslov (наслов) titula (титула) natpis (натпис)	nadpis titul názov	titel
hasło tytułowa	entrada pelo título	catalogizaţie cu titlu	stvarni naslov (стварни наслов)	názov	titelhänvisning
karta tytułowa	rosto	frontispiciu	naslovna strana (насловна страна)	titulný list	titelblad
do	a para	la	do (до) k (к) ka (ка)	k ku	för till
tragedia	tragédia	tragedie	tragedija (трагедија)	tragédia	sorgespel tragedi
wiadomości *pl.* acta *pl.*	comunicado	buletin	izveštaj (извештај) saopštenja (саопштења) *pl.* bilten (билтен)	správy *pl.* oznámenie oznam	meddelande bulletin skrifter *pl.*
transkrypcja	transcrição	transcripţie transcriere	transkripcija (транскрипција)	transkripcia	transkription
przetłumaczyć przełożyć	traduzir verter	traduce	prevoditi (преводити) tumačiti (тумачити)	prekladať	översätta

	ENGLISH Anglica	**FRENCH** Gallica	**GERMAN** Germanica	**RUSSIAN** Russica	**SPANISH** Hispanica
349	translation	traduction	Übersetzung	perevod (перевод)	traducción
350	transliteration	translitération	Transliteration	transliteracija (транслитерация)	transliteración
351	treat (discuss)	traiter	behandeln	rassmatrivať (рассматривать) razrabatyvat' (разрабатывать) traktovat' (трактовать)	tratar disertar
352	treatise (essay[2])	traité dissertation	Abhandlung Traktat	naučnyj traktat (научный трактат) trud (труд)	tratado disertación
353	typescript (typed copy, typewritten manuscript)	texte dactylographié	Schreibmaschinen- abschrift maschinengeschrie- benes Manuskript Typoskript	mašinopis' (машинопись)	ejemplar mecano- grafiado
354	type size (body size)	corps (des caractères) force de corps	Schriftgrad Typengröße	kegľ šrifta (кегль шрифта)	fuerza del cuerpo
	typewritten manu- script →typescript				
	typographical error →printer's error				
355	typography	typographie	Typographie	knigopečatanie (книгопечатание) poligrafija (полиграфия)	tipografía
	unaltered →unchanged				
356	unchanged (unaltered)	inchangé invariable	unverändert	bez izmenenij (без изменений) neizmennyj (неизменный)	invariable
357	under	sous	unter	pod (под)	bajo debajo de

BULGARIAN Bulgarica	CROATIAN Croatica	CZECH Bohemica	DANISH Danica	DUTCH Hollandica	FINNISH Fennica
prevod (превод)	prijevod	překládání překlad	oversættelse	vertaling	käännös
transliteracija (транслитерация)	transliteracija	transliterace	translitteration	translitteratie	translitteraatio translitterointi
tretiram (третирам) razgležǎam (разглеждам)	raspravljati	pojednávat	behandle	behandelen	keskustella
traktat (трактат) teza (теза)	rasprava	pojednání traktát rozprava	afhandling traktat	verhandeling tractaat	keskustelu tutkielma väitöskirja
mašinopisen ekzempljar (машинописен екземпляр)	prijepis na pisa- ćem stroju	strojopis	maskinskrevet eksemplar/manu- skript	maschineschrift	koneella kirjoitettu kappale konekirjoite
golemina na šrifta (големина на шрифта)	vrsta pisma	kuželka	skriftgrad	korps	kirjasinaste
tipografija (типография) poligrafija (полиграфия)	tiskara	knihtisk polygrafie	typografi	drukkerij typografie	typografia
nesmenen (несменен) neizmenen (неизменен)	nepromjenjiv nepromjenljiv	nezměněný	uforandret	onveranderd	muuttamaton
pod (под)	ispod pod	pod	under	onder	alla alle

	GREEK Graeca	HUNGARIAN Hungarica	ITALIAN Italica	LATIN Latina	NORWEGIAN Norvegica
349	metaphrasis (μετάφρασις)	fordítás	traduzione	conversio translatio interpretatio	oversettelse
350	metagraphe (μεταγραφή)	betű szerinti átírás	traslitterazione	translitteratio	translitterasjon
351	pragmateuomai (πραγματεύομαι)	értekezik tárgyal	dissertare trattare	dissertare tractare consiliari	behandle traktere
352	diatribe (διατριβή) pragmateia (πραγματεία)	értekezés tárgyalás	trattato	dissertatio tractatio tractatulus	avhandling disputas
353	daktilographemenon (δακτιλογραφημένον)	gépelt kézirat gépirat	dattiloscritto	dactylographia machina scriptum	maskinskrevet eksemplar/manu- skript
354	eidos stoicheion (εἶδος στοιχείων)	betűnagyság	corpo tipografico	gradus typorum	skriftgrad
355	typographia (τυπογραφία)	nyomdászat tipográfia	tipografia	typographia	typografi
356	ametabletos (ἀμετάβλητος)	változatlan	invariato	immutabilis immutatus	uforanderlig uforandret
357	hypo (ὑπό)	alatt alá	sotto	sub subter	under

POLISH Polonica	PORTUGUESE Portugallica	RUMANIAN Rumenica	SERBIAN Servica	SLOVAK Slovaca	SWEDISH Suecica
przetłumaczenie przekład	tradução	traducere	prevod (превод)	preklad	översättning
transliteracja	transliteração	transliterație	transliteracija (транслитерација)	transliterácia	translitteration
rozprawiać	dissertar tratar	diserta	raspravljati (расправљати)	pojednávať rokovať	avhandla behandla
rozprawa traktat	dissertação	disertație tratat	rasprava (расправа) traktat (трактат)	pojednanie vedecká rozprava	avhandling traktat
maszynopis	exemplar dactilografado	dactilografiat	daktilografisani primerak (дактилографи- сани примерак)	strojopis strojopisný exemplár	maskinskrift
stopień czcionki	corpo do tipo	caracter	tipovi slova (типови слова)	veľkosť písmena	typgrad
drukarnia typografia	tipografia	tipografie	štamparstvo (штампарство) tiskara (тискара)	tlačiarstvo polygrafia	typografi
niezmieniony	inalterável invariável	invariabil neschimbat	nepromenljiv (непроменљив)	nezmenený bezo zmeny	oförändrad
pod	debaixo de por baixo de sob	sub	ispod (испод) pod (под)	pod	under

		ENGLISH Anglica	FRENCH Gallica	GERMAN Germanica	RUSSIAN Russica	SPANISH Hispanica
358		union catalog(ue)	catalogue collectif	Gesamtkatalog Zentralkatalog	svodnyj katalog (сводный каталог)	catálogo colectivo
		unique →only				
359		university	université	Universität	universitet (университет)	universidad
360		university library	bibliothèque universitaire	Universitäts- bibliothek	universitetskaja biblioteka (университетская библиотека)	biblioteca universitaria
361		unpublished	inédit	unveröffentlicht ungedruckt	neopublikovannyj (неопубликованный) neizdannyj (неизданный)	inedito
362		use n.	usage emploi	Gebrauch Benutzung	polzovanie (пользование) primenenie (применение)	uso empleo
363		use v. (employ)	user (de) employer	gebrauchen	ispolzovať (использовать) primenjať (применять)	usar emplear
		variant →version				
		verse →poem				
364		version (variant)	variante version	verschiedene Lesart Fassung Version	versija (версия) variant (вариант) raznočtenie (разночтение)	versión variante
365		verso (left-hand page)	verso	verso Rückseite	oborotnaja storona lista (оборотная сторона листа)	verso
		vocabulary →glossary				
366		volume[1] (tome)	volume tome	Band	tom (том)	volumen tomo

BULGARIAN Bulgarica	CROATIAN Croatica	CZECH Bohemica	DANISH Danica	DUTCH Hollandica	FINNISH Fennica
svoden katalog (своден каталог)	centralni katalog	souborný katalog	fælleskatalog	centrale catalogus	yhteisluettelo
universitet (университет)	sveučilište	universita	universitet	universiteit	yliopisto
universitetska biblioteka (университетска библиотека)	sveučilišna knjižnica	universitní knihovna	universitets- bibliotek	universiteits- bibliotheek	yliopisto(n)- kirjasto
neizdaden (неиздаден)	neizdan	nevydaný	utrykt ikke udgivet	onuitgegeven ongedrukt	julkaisematon
polzuvane (ползуване) upotreba (употреба)	upotreba korišćenje	upotřebení užití	anvendelse brug	aanwending gebruik	käyttö käytäntö
upotrebjavam (употребявам)	upotrebljavati koristiti	upotřebit užít užívat	anvende bruge	gebruiken	käyttää
variant (вариант) versija (версия)	varijant verzija	variace změna verse	variant version	variant versie	muunnos
obratna strana (обратна страна)	naličje	rub verso	bagside	keerzijde verso	kääntöpuoli lehden takasivu selkäpuoli
tom (том)	svezak	svazek díl	bind	band	nidos nide

	GREEK Graeca	HUNGARIAN Hungarica	ITALIAN Italica	LATIN Latina	NORWEGIAN Norvegica
358	synkentrotikos katalogos (συγκεντρωτικός κατάλογος)	központi katalógus	catalogo collettivo	catalogus centralis	felleskatalog
359	panepistemion (πανεπιστήμιον)	egyetem	università	universitas (scientiarum)	universitet
360	panepistemiake bibliotheke (πανεπιστημιακή βιβλιοθήκη)	egyetemi könyvtár	biblioteca universitaria	bibliotheca universitatis (scientiarum)	universitetsbibliotek
361	anekdotos (ἀνέκδοτος)	kiadatlan	inedito	ineditus	ikke offentliggjort ikke udgivet utrykt
362	chresis (χρῆσις)	használat	uso impiego	usus	anvendelse bruk
363	chresimopoio (χρησιμοποιῶ)	használ	usare	uti	anvende bruke
364	metabole (μεταβολή) metastasis (μεταστάσις)	változat	variante versione	varians variatio	variasjon
365	hetera pleura (ἕτερα πλεῦρα)	verzó	verso	verso	bakside
366	tomos (τόμος)	kötet	volume tomo	tomus volumen liber	bind

POLISH Polonica	PORTUGUESE Portugallica	RUMANIAN Rumenica	SERBIAN Servica	SLOVAK Slovaca	SWEDISH Suecica
centralny katalog	catálogo colectívo	catalog central	centralni katalog (централни каталог)	ústredný katalóg	centralkatalog
uniwersytet	universidade	universitate	univerzitet (универзитет)	univerzita	universitet
biblioteka uniwersytecka	biblioteca universitária	bibliotecă universitară	univerzitetska biblioteka (универзитетска библиотека)	univerzitná knižnica	universitets- bibliotek
nie drukowany nie wydany	inédito não publicado	inedit	neizdan (неиздан)	nevydaný	outgiven otryckt
użycie użytek	emprêgo uso	întrebuinţare uz	upotreba (употреба) korišćenje (коришћење)	upotrebenie užívanie	bruk användning
użytkować	usar	întrebuinţa uza	upotrebljavati (употребљавати) koristiti (користити)	upotrebiť užívať	använda bruka
odmiana wariant wersja	variante versão	variantă versiune	varijant (варијант) verzija (верзија)	variácia variant obmena	variant version
odwrocie strona odwrotna	verso	contrapagina verso	naličje (наличје)	druhá strana rub	baksida verso
tom	tomo volume	volum tom	tom (том) svezak (свезак) sveska (свеска)	zväzok	volym band

	ENGLISH Anglica	FRENCH Gallica	GERMAN Germanica	RUSSIAN Russica	SPANISH Hispanica
367	volume² (annual volume)	année	Jahrgang	godovoj komplekt (годовой комплект)	año
	whole →complete				
368	with	avec	mit	s, so (с, со)	con
369	without	sans	ohne	bez (без)	sin
370	woodcut (wood engraving)	bois gravure sur bois xylographie	Holzschnitt Holztafeldruck	gravjura na dereve (гравюра на дереве)	grabado en madera xilografía
	words *pl* →text				
371	work	œuvre ouvrage	Werk	trud (труд) proizvedenie (произведение) sočinenie (сочинение)	obra
	work out (in detail) →elaborate				
372	write	écrire	schreiben	pisať (писать)	escribir
373	write poetry	composer/ écrire des vers	dichten	pisať/sočinjať stichi (писать/сочинять стихи)	componer (versos)
374	writer	écrivain	Schriftsteller	pisateľ (писатель)	escritor
	writing for the blind →braille printing				
375	xerography	xérographie	Xerographie	kserografija (ксерография)	jerografía xerografía
376	year	an année	Jahr	god (год)	año
377	year-book (annual)	annuaire	Jahrbuch (*pl.* Jahrbücher)	ežegodnik (ежегодник)	anuario

BULGARIAN Bulgarica	CROATIAN Croatica	CZECH Bohemica	DANISH Danica	DUTCH Hollandica	FINNISH Fennica
godišnina (годишнина) godina (година)	godište	ročník rok	årgang	jaargang	vuosikerta
s (c) sås (със)	s sa	s	med	door met	kanssa mukana kera -lla -llä
bez (без)	bez	bez	uden	zonder	ilman
gravjura na dârvo (гравюра на дърво)	drvorez	dřevoryt	træsnit træstik	houtsnede	puupiirros
trud (труд) proizvedenie (произведение) sâčinenie (съчинение)	djelo tvorevina	dílo	værk arbejde	werk	teos
piša (пиша)	pisati	psát	skrive	schrijven	kirjoittaa
piša stihove (пиша стихове) sâčinjavam (съчинявам)	pisati pjesme	psát verše/básně	digte	dichten	runoilla
pisatel (писател)	pisac književnik	spisovatel	forfatter	schrijver	kirjailija tekijä
kserografija (ксерография)	kserografija	xerografie	xerografi	xerografie	kseropainanta
godina (година)	godina ljeto	rok	år	jaar	vuosi
godišnik (годишник)	godišnjak	ročenka	årbog	jaarboek	vuosikirja

	GREEK Graeca	**HUNGARIAN** Hungarica	**ITALIAN** Italica	**LATIN** Latina	**NORWEGIAN** Norvegica
367	ep'eniauton ephemerides (ἐπ' ἐνιαυτόν ἐφημερίδες) epeteios (ἐπέτειος)	évfolyam	annata	cursus annuus annus	årgang
368	meta (μετά) me (μέ) syn (σύν)	-val, -vel	con	cum	med ved
369	aneu (ἄνευ)	nélkül	senza	sine	uten
370	xylographia (ξυλογράφια)	fametszet	incisione in legno xilografia	imago in ligno incisa xylographia	tresnitt
371	ergon (ἔργον) syngraphe (συγγραφή)	mű	opera (collettiva)	opus(culum) elaboratum	arbeid verk
372	grapho (γράφω) syngrapho (συγγράφω)	ír	scrivere	conscribere scribere	skrive
373	grapho poiemata (γράφω ποιήματα)	verset ír költ	comporre versi	versus facere componere	dikte
374	syngrapheus (συγγραφεύς)	író	scrittore	scriptor	forfatter
375	xerographia (ξηρογραφία)	xerográfia	xerografia	xerographia	xerografi
376	etos (ἔτος) chronos (χρόνος)	év esztendő	anno	annus	år
377	epeteris (ἐπετηρίς)	évkönyv	annuario	liber annalis	årbok

POLISH Polonica	PORTUGUESE Portugallica	RUMANIAN Rumenica	SERBIAN Servica	SLOVAK Slovaca	SWEDISH Suecica
rocznik	um ano	colecţie anuală an	godište (годиште)	ročník	årgång
z(e)	com junto	cu	s (c) sa (ca)	s so	med
bez	sem	fără	bez (без)	bez	utan
drzeworyt	gravura em madeira	gravură în lemn xilogravură	drvorez (дрворез)	drevorezba drevoryt	träsnitt
dzieło utwór	obra trabalho	operă	delo (дело) tvorevina (творевина)	dielo	verk arbete
pisać	escrever	scrie	pisati (писати)	písať	skriva
pisać wiersze	fazer versos	compune versuri	pisati pesme (писати песме)	skadať básne	skriva vers
pisarz	escritor	scriitor	pisac (писац) književnik (књижевник)	spisovateľ	skriftställare
kserografia	xerografia	xerografie	kserografija (ксерографија)	xerografia	xerografi
rok	ano	an	godina (година) leto (лето)	rok	år
rocznik	anuário	anuar	godišnjak (годишњак)	letopis ročenka	årsbok

	ENGLISH Anglica	**FRENCH** Gallica	**GERMAN** Germanica	**RUSSIAN** Russica	**SPANISH** Hispanica
					APPENDIX
I	**January**	janvier	Januar Jänner	janvar' (январь)	enero
II	**February**	février	Februar Feber	fevral' (февраль)	febrero
III	**March**	mars	März	mart (март)	marzo
IV	**April**	avril	April	aprel' (апрель)	abril
V	**May**	mai	Mai	maj (май)	mayo
VI	**June**	juin	Juni	ijun' (июнь)	junio
VII	**July**	juillet	Juli	ijul' (июль)	julio
VIII	**August**	août	August	avgust (август)	agosto
IX	**September**	septembre	September	sentjabr' (сентябрь)	septiembre
X	**October**	octobre	Oktober	oktjabr' (октябрь)	octubre
XI	**November**	novembre	November	nojabr' (ноябрь)	noviembre

BULGARIAN Bulgarica	CROATIAN Croatica	CZECH Bohemica	DANISH Danica	DUTCH Hollandica	FINNISH Fennica
APPENDIX					
januari (януари)	januar siječanj	leden	januar	januari	tammikuu
fevruari (февруари)	februar veljača	únor	februar	februari	helmikuu
mart (март)	mart ožujak	březen	marts	maart	maaliskuu
april (април)	april travanj	duben	april	april	huhtikuu
maj (май)	maj svibanj	květen	maj	mei	toukokuu
juni (юни)	junije lipanj	červen	juni	juni	kesäkuu
juli (юли)	julije spranj	červenec	juli	juli	heinäkuu
avgust (август)	august kolovoz	srpen	august	augustus	elokuu
septemvri (септември)	septembar rujan	září	september	september	syyskuu
oktomvri (октомври)	oktobar listopad	říjen	oktober	october	lokakuu
noemvri (ноември)	novembar studeni	listopad	november	november	marraskuu

		GREEK Graeca	HUNGARIAN Hungarica	ITALIAN Italica	LATIN Latina	NORWEGIAN Norvegica
I		Ianuarios (’Ιανουάριος) Gamelion (Γαμηλιών)	január	gennaio	Ianuarius	januar
II		Februarios (Φεβρουάριος) Anthesterion (’Ανθεστηριών)	február	febbraio	Februarius	februar
III		Martios (Μάρτιος) Elaphebolion (’Ελαφηβολιών)	március	marzo	Martius	mars
IV		Aprilios (’Απρίλιος) Munichion (Μουνιχιών)	április	aprile	Aprilis	april
V		Maios (Μάϊος) Thargelion (Θαργηλιών)	május	maggio	Maius	mai
VI		Iunios (’Ιούνιος) Skirophorion (Σκιροφοριών)	június	giugno	Iunius	juni
VII		Iulios (’Ιούλιος) Hekatonbaion (’Εκατονβαιών)	július	luglio	Iulius	juli
VIII		Augustos (Αὔγουστος) Metageitnion (Μεταγειτνιών)	augusztus	agosto	Augustus	august
IX		Septembrios (Σεπτέμβριος) Boedromion (Βοηδρομιών)	szeptember	settembre	September	september
X		Oktobrios (’Οκτώβριος) Pyanopsion (Πυανοψιών)	október	ottobre	October	oktober
XI		Noembrios (Νοέμβριος) Maimakterion (Μαιμακτηριών)	november	novembre	November	november

POLISH Polinica	PORTUGUESE Portugallica	RUMANIAN Rumenica	SERBIAN Servica	SLOVAK Slovaca	SWEDISH Suecica
styczeń	Janeiro	ianuarie	januar (јануар)	január	januari
luty	Fevereiro	februarie	februar (фебруар)	február	februari
marzec	Março	martie	mart (март)	marec	mars
kwiecień	Abril	aprilie	april (април)	apríl	april
maj	Maio	mai	maj (мај)	máj	maj
czerwiec	Junho	iunie	juni (јуни)	jún	juni
lipiec	Julho	iulie	juli (јули)	júl	juli
sierpień	Agosto	august	avgust (август)	august	augusti
wrzesień	Setembro	septemvrie	septembar (септембар)	september	september
październik	Outubro	octomvrie	oktobar (октобар)	október	oktober
listopad	Novembro	noiembrie	novembar (новембар)	november	november

	ENGLISH Anglica	FRENCH Gallica	GERMAN Germanica	RUSSIAN Russica	SPANISH Hispanica
XII	**December**	décembre	Dezember	dekabr' (декабрь)	diciembre
XIII	**Monday**	lundi	Montag	ponedel'nik (понедельник)	lunes
XIV	**Tuesday**	mardi	Dienstag	vtornik (вторник)	martes
XV	**Wednesday**	mercredi	Mittwoch	sreda (среда)	miércoles
XVI	**Thursday**	jeudi	Donnerstag	četverg (четверг)	jueves
XVII	**Friday**	vendredi	Freitag	pjatnica (пятница)	viernes
XVIII	**Saturday**	samedi	Samstag Sonnabend	subbota (суббота)	sábado
XIX	**Sunday**	dimanche	Sonntag	voskresen'e (воскресенье)	domingo
XX	**year**	an année	Jahr	god (год)	año
XXI	**annual** **yearly**	annuel	jährlich	ežegodnyj (ежегодный) godovoj (годовой)	anual
XXII	**half-yearly** **semi-annual**	semestriel	halbjährlich	polugodovoj (полугодовой) šestimesjačnyj (шестимесячный)	semestral
XIII	**quarterly**	′trimestriel	dreimonatlich vierteljährlich	trëhmesjačnyj (трёхмесячный)	trimestral
XXIV	**month**	mois	Monat	mesjac (месяц)	mes
XXV	**monthly**	mensuel	monatlich	ežemesjačnyj (ежемесячный)	mensual
XXVI	**fortnightly** **semi-monthly**	bimensuel semi-mensuel	halbmonatlich	polumesjačnyj (полумесячный)	semimensual
XXVII	**week**	semaine	Woche	nedelja (неделя)	semana

BULGARIAN Bulgarica	CROATIAN Croatica	CZECH Bohemica	DANISH Danica	DUTCH Hollandica	FINNISH Fennica
dekemvri (декември)	decembar prosinac	prosinec	december	december	joulukuu
ponedelnik (понеделник)	ponedjeljak	pondělí	mandag	maandag	maanantai
vtornik (вторник)	utornik	úterý	tirsdag	dinsdag	tiistai
srjada (сряда)	srijeda	středa	onsdag	woensdag	keskiviikko
četvârtâk (четвъртък)	četvrtak	čtvrtek	tørsdag	donderdag	torstai
petâk (петък)	petak	pátek	fredag	vrijdag	perjantai
sâbota) (събота)	subota	sobota	lørdag	zaterdag	lauantai
nedelja (неделя)	nedelja	neděle	søndag	zondag	sunnuntai
godina (година)	godina	rok	år	jaar	vuosi
godišen (годишен)	godišnji	roční	årlig	jaarlijks	vuotinen
polugodišen (полугодишен)	polugodišnji	šestiměsíční	halvårlig	halfjaarlijks	puolivuotinen
trimesečen (тримесечен)	tromjesečni	tříměsíční	kvartals	driemaandelijks	kolmen kuukauden
mesec (месец)	mjesec	měsíc	måned	maand	kuukausi
mesečen (месечен)	mjesečni	měsíční	månedlig	maandelijks	jokakuukautinen
polumesečen (полумесечен)	polumjesečni	čtrnáctidenní	halvmånedlig	halfmaandelijks	puolikuukautinen
sedmica (седмица)	tjedan sedmica	týden	uge	week	viikko

	GREEK Graeca	HUNGARIAN Hungarica	ITALIAN Italica	LATIN Latina	NORWEGIAN Norvegica
XII	Dekembrios (Δεκέμβριος) Poseideon (Ποσειδεών)	december	dicembre	December	desember
XIII	Deutera (Δευτέρα)	hétfő	lunedì	Dies Lunae feria II.	mandag
XIV	Trite (Τρίτη)	kedd	martedì	Dies Martis feria III.	tirsdag
XV	Tetarte (Τετάρτη)	szerda	mercoledì	Dies Mercurii feria IV.	onsdag
XVI	Pempte (Πέμπτη)	csütörtök	giovedì	Dies Jovis feria V.	torsdag
XVII	Paraskeue (Παρασκευή)	péntek	venerdì	Dies Veneris feria VI.	fredag
XVIII	Sabbaton (Σάββατον)	szombat	sabato	Dies Saturni feria VII.	lørdag
XIX	Kyriake (Κυριακή)	vasárnap	domenica	Dies Solis Dominica	søndag
XX	etos (ἔτος)	év	anno	annus	år
XXI	etesios (ἐτήσιος)	évi	annuale	annalis annuus	årlig
XXII	hexamerios (ἑξάμηνος)	félévi	semestrale	semestris	halvårlig
XXIII	trimeniaios (τριμηνιαῖος)	negyedévi	trimestrale	trimestris	kvartal
XXIV	men (μήν)	hó hónap	mese	mensis	måned
XXV	meniaios (μηνιαῖος)	havi	mensile	mensilis menstruus	månedlig
XXVI	dekapenthemeros (δεκαπενθήμερος)	félhavi	semimensile	semimensilis semimenstruus	halvmånedlig toukentlig
XXVII	hebdomas (ἑβδομάς)	hét	settimana	hebdomas hebdomada	uke

POLISH Polonica	PORTUGUESE Portugallica	RUMANIAN Rumenica	SERBIAN Servica	SLOVAK Slovaca	SWEDISH Suecica
grudzień	Dezembro	decembrie	decembar (децембар)	december	decemrbe
poniedziałek	segunda-feira	luni	ponedeljak (понедељак)	pondelok	mandag
wtorek	terça-feira	marţi	utorak (уторак)	útorok	tisdag
środa	quarta-feira	mercuri	sreda (среда)	streda	onsdag
czwartek	quinta-feira	joi	četvrtak (четвртак)	štvrtok	torsdag
piątek	sexta-feira	vineri	petak (петак)	piatok	fredag
sobota	sábado	sîmbătă	subota (субота)	sobota	lördag
niedziela	domingo	duminică	nedelja (недеља)	nedeľa	söndag
rok	ano	an	godina (година)	rok	år
roczny	anual	anual	godišnji (годишњи)	ročný	årlig
półroczny	semestral	bianual semestrial	polugodišnji (полугодишњи)	polročný	halvårs-
kwartalny	trimensal trimestral	trimestrial	tromesečni (тромесечни)	štvrťročný	kvartals-
miesiąc	mês	lună	mesec (месец)	mesiac	månad
miesięczny	mensal	lunar	mesečni (месечни)	mesačný	månatlig
półmiesięczny	bimensal quinzenal	bilunar	polumesečni (полумесечни)	polmesačný	halvmånads-
tydzień	semana	săptămînă	nedelja (недеља) sedmica (седмица)	týždeň	vecka

185

	ENGLISH Anglica	FRENCH Gallica	GERMAN Germanica	RUSSIAN Russica	SPANISH Hispanica
XXVIII	weekly	hebdomadaire	wöchentlich	eženedeľnyj (еженедельный)	semanal
XXIX	day	jour	Tag	den' (день)	día
XXX	daily	quotidien	täglich	eždnevnyj (ежедневный)	cotidiano diario

BULGARIAN Bulgarica	CROATIAN Croatica	CZECH Bohemica	DANISH Danica	DUTCH Hollandica	FINNISH Fennica
sedmičen (седмичен)	tjedni sedmični	týdenní	ugentlig	wekelijks	viikko-
den (ден)	dan	den	dag	dag	päivä
vsekidneven (всекидневен)	dnevni	denní	daglig	dagelijks	päivä-

	GREEK Graeca	HUNGARIAN Hungarica	ITALIAN Italica	LATIN Latina	NORWEGIAN Norvegica
XXVIII	hebdomadiaios (ἑβδομαδιαῖος)	heti	settimanale	semel in hebdomade	ukentlig
XXIX	(he)mera (ἡμέρα)	nap	giorno	dies	dag
XXX	hemeresios (ἡμερήσιος)	napi	quotidiano	cotidianus	daglig

POLISH Polonica	PORTUGUESE Portugallica	RUMANIAN Rumenica	SERBIAN Servica	SLOVAK Slovaca	SWEDISH Suecica
tygodniowy	semanal	săptămînal ebdomadar	sedmični (седмични)	týždňový	vecko-
dzień	dia	zi	dan (дан)	deň	dag
codzienny	diário quotidiano	cotidian zilnic	dnevni (дневни)	denný dňový	daglig

II

FRENCH — GALLICA

bibliothèque de référence, reference library	279
bibliothèque d'État, state library	322
bibliothèque municipale, city library	68
bibliothèque nationale, national library	207
bibliothèque publique, public library	270
bibliothèque spécialisée, special library	318
bibliothèque universitaire, university library	360
bilingue, bilingual	44
biobibliographie, author bibliography	36
biographie, biography	48
bois, woodcut	370
brocher, sew	306
brochure, booklet; number[2]	50; 215
bulletin, bulletin; report[1]	54; 283
bulletin (d'informations), transactions *pl.*	346
bureau, office	219
cahier, number[2]	215
carte géographique, map	193
catalogage, cataloging	60
catalogue, catalog(ue) *n.*	58
catalogue collectif, union catalog(ue)	358
cataloguement, cataloging	60
cataloguer, catalog(ue) *v.*	59
catalogue sur fiches, card index	57
centre des échanges, exchange centre	138
chalcotypie, copperplate	94
changer, change[1]; change[2]	63; 64
chant, song	314
chant nuptial, bridal/nuptial song	53
chapitre, chapter	65
chef, chief	66
chiffre, number[1]	214
choisir, select	302
chrestomathie, selection	303
chronique, review	287
classement alphabétique, alphabetical order	15
classification, classification	69
classification décimale, decimal classification	107
clause, clause	70
co-auteur, joint author	174
codex, codex	71
collaborateur, collaborator	73
collaborer, collaborate	72
collecter, collect	74
collection, collection	75
collectivité-auteur, corporate author	100
collectivité d'auteurs, corporate author	100

colonne, column	77
comédie, comedy	78
comité, committee	83
commande, order[2]	225
commentaire, commentary	82
commenter, comment[1] *v.*	81
commission, committee	83
compiler, compile	84
complet, complete	85
compléter, complete *v.*	87
composer, compile	84
composer des vers write poetry	373
composition, matter	195
comprendre, contain	90
concourir, collaborate	72
conférence, conference; lecture	88; 179
congrès, congress	89
conte, short story, story	309; 324
conte (de fées), fable	143
contenir, contain	90
contenu, contents *pl.*	91
continuer, continue	92
contribution, contribution	93
coopérer, collaborate	72
copie, copy[1]; imitation	96; 161
corps (des caractères), type size	354
correction, correction	102
correspondance, correspondence	103
corriger, correct	101
cote, location mark	189
coudre, sew	306
critique, critique	104
critiquer, criticize	105
croquis, sketch	312
d', about; from; of	3; 153; 218
dans, in	164
date d'impression, date of printing	106
de, about; from; of; on	3; 153; 218; 222
dédicacé, dedicated	108
dédicace, dedication	109
dédié, dedicated	108
défectueux, defective	110
défendre, defend	111
dépôt de livres, stack-room	320
dépôt des actes, archives *pl.*	28
dépôt légal, copyright deposit	99
desiderata *pl.*, desiderata *pl.*	113

reproduction, reprint[1]; reproduction	285; 286
résumé, summary	330
résumer, summarize	329
retravailler, rewrite	291
reviser, revise[1]	289
révision, revision	290
revoir, revise[1]	289
revue, review	287
roman, novel	213
sans, without	369
scénario, scenario	294
science-fiction, science fiction	297
sciences pl., science	296
séance, sitting	311
section, section	301
sélectionner, select	302
série, series	305
service des échanges, exchange centre	138
siècle, century	62
signature, signature	310
s. l. n. d. → sans lieu ni date	
société, society	313
sommaire, contents pl.; summary; table of contents	91; 330; 334
source, source	316
sous, under	357
statuts pl. statute[1]	323
sténographie, shorthand	308
strophe, strophe	325
suite, sequel; series	304; 305
suppléer, complete v.	86
supplément, supplement	331
supplémentaire, supplementary	332
sur, about; on	3; 222
tableau, table	333
table des matières, table of contents	334
taille-douce, copperplate engraving	95
texte, text	337
texte dactylographié, typescript	353
thèse, thesis	340
tirage à part, offprint	221
tiré à part, offprint	221
tirer, print[1]	261
titre, title	341
titre courant, running title	292
tome, volume[1]	366
traduction, translation	349
traduire, translate	348
tragédie, tragedy	345
traité, treatise	352
traiter, treat	351
trame, screen	298
transcription, transcription	347
translitération, transliteration	350
typographie, typographpy	355
un, une a, an	1
unique, only	223
université, university	359
usage, use	362
user (de), use	363
variante, version	364
vers, poem	248
version, version	364
verso, verso	365
vieux, ancient	17
vocabulaire, dictionary; glossary	116; 154
volume, volume[1]	366
xérographie, xerography	375
xylographie, woodcut	370

APPENDIX

CALENDAR UNITS — TEMPORA

janvier, January	I	**décembre,** December	XII	**trimestriel,** quarterly	XXIII
février, February	II	**lundi,** Monday	XIII	**mois,** month	XXIV
mars, March	III	**mardi,** Tuesday	XIV	**mensuel,** monthly	XXV
avril, April	IV	**mercredi,** Wednesday	XV	**bimensuel, semimensuel,**	
mai, May	V	**jeudi,** Thursday	XVI	fortnightly, semi-	
juin, June	VI	**vendredi,** Friday	XVII	monthly	XXVI
juillet, July	VII	**samedi,** Saturday	XVIII	**semaine,** week	XXVII
août, August	VIII	**dimanche,** Sunday	XIX	**hebdomadaire,** weekly	XXVIII
septembre, September	IX	**an, année,** year	XX	**jour,** day	XXIX
octobre, October	X	**annuel,** annual, yearly	XXI	**quotidien,** daily	XXX
novembre, November	XI	**semestriel,** half-yearly,			
		semi-annual	XXII		

NUMERALS — NUMERI

1 **un, une**	one	15 **quinze**	fifteen
2 **deux**	two	16 **seize**	sixteen
3 **trois**	three	17 **dix-sept**	seventeen
4 **quatre**	four	18 **dix-huit**	eighteen
5 **cinq**	five	19 **dix-neuf**	nineteen
6 **six**	six	20 **vingt**	twenty
7 **sept**	seven	21 **vingt et un**	twenty one
8 **huit**	eight	30 **trente**	thirty
9 **neuf**	nine	40 **quarante**	forty
10 **dix**	ten	50 **cinquante**	fifty
11 **onze**	eleven	60 **soixante**	sixty
12 **douze**	twelve	70 **soixante-dix**	seventy
13 **treize**	thirteen	80 **quatre-vingt(s)**	eighty
14 **quatorze**	fourteen	90 **quatre-vingt-dix**	ninety

100 cent	hundred	15. quinzième	fifteenth
200 deux cents	two hundred	16. seizième	sixteenth
300 trois cents	three hundred	17. dix-septième	seventeenth
400 quatre cents	four hundred	18. dix-huitième	eighteenth
500 cinq cents	five hundred	19. dix-neuvième	nineteenth
600 six cents	six hundred	20. vingtième	twentieth
700 sept cents	seven hundred	21. vingt et unième	twenty-first
800 huit cents	eight hundred	30. trentième	thirtieth
900 neuf cents	nine hundred	40. quarantième	fortieth
1000 mille	thousand	50. cinquantième	fiftieth
1. premier, -ère	first	60. soixantième	sixtieth
2. deuxième, second, -e	second	70. soixante-dixième	seventieth
3. troisième	third	80. quatre-vingtième	eightieth
4. quatrième	fourth	90. quatre-vingt-dixième	ninetieth
5. cinquième	fifth	100. centième	hundredth
6. sixième	sixth	200. deux centième	two hundredth
7. septième	seventh	300. trois centième	three hundredth
8. huitième	eighth	400. quatre centième	four hundredth
9. neuvième	ninth	500. cinq centième	five hundredth
10. dixième	tenth	600. six centième	six hundredth
11. onzième	eleventh	700. sept centième	seven hundredth
12. douzième	twelfth	800. huit centième	eight hundredth
13. treizième	thirteenth	900. neuf centième	nine hundredth
14. quatorzième	fourteenth	1000. millième	thousandth

GERMAN — GERMANICA

Band, volume[1]	366	Buchwesen, bibliology		43
bearbeiten, adapt; edit[1]	8; 122	Bühnenwerk, play		246
beglaubigt, authentic	34	Bulletin, bulletin		54
behandeln, treat	351	Büro, office		219
Beiblatt, supplement	331			
Beilage, supplement	331	Comics, comic strip		80
beilegen, annex	20	Comic strip, comic strip		80
beischließen, annex	20			
Beitrag, contribution	93	das, the		339
Belletristik, fiction	146	defekt, defective; incomplete	110;	165
Bemerkung, note[1]	212	Denkschrift, memoir		196
Benutzung, use	362	der, the		339
berechtigen, authorize	37	Desiderat(um), desiderata *pl.*		113
Bericht, bulletin; report[1]	54; 283	Dezimalklassifikation, decimal classification		107
besorgen (die Ausgabe), edit[1]	122	Dialog, dialog(ue)		114
besprechen, criticize; review *v.*	105; 288	dichten, write poetry		373
Bestand, inventory	173	Dichter, poet		249
Bestellung, order[2]	225	Dichtung, poem		248
Bezeichnungsweise, notation	211	die, the		339
Bibliographie, bibliography	42	Dissertation, thesis		340
Bibliothek, library	183	Dokumentation, documentation		118
Bibliothekar, librarian	182	Doppelstück, duplicate		121
Bild, picture	241	Drama, play		246
Bilddruck, engraving	132	Drehbuch, scenario		294
Bildnis, portrait	253	drucken, print[1]		261
Bildtafel, plate	245	Drucker, printer		264
binden, bind	45	Druckerei, printing office		266
Biobibliographie, author bibliography	36	Druckerwerkstatt, printing office		266
Biographie, biography	48	Druckfehler, printer's error		265
bis, to	344	Druckjahr, date of printing		106
Blatt, leaf	178	Drucksache, printed matter		262
Blattrand, margin	194	Druckvermerk, imprint		163
Blindenschrift, braille printing	52	Dublette, duplicate		121
Bogen, sheet	307			
Brief, letter	181	echt, authentic		34
Briefwechsel, correspondence	103	ein, eine, ein, a, an		1
broschiertes Buch, paperback	232	Einband, binding		47
Broschüre, booklet	50	einfügen, insert		169
Bruchstück, fragment	152	einführen, introduce		171
Buch, book	49	Einführung, introduction		172
Buchbinderei, bindery	46	eingegangen, ceased publication		61
Buchdrucker, printer	264	einleiten, introduce		171
Bücherkunde, bibliography	42	Einleitung, introduction		172
Büchermagazin, stack-room	320	Eintragung unter dem Sachtitel, title entry		342
Bücherspeicher, stack-room	320	einzig, only *a.*		223
Buchhandlung, bookshop	51	Elegie, elegy		129
Buchkunde, bibliology	43	enthalten, contain		90
Buchladen, bookshop	51	Entwurf, sketch		312

Enzyklopädie, encyclopaedia	130
Epilog, epilogue	135
Epistel, letter	181
Epos, epic	134
ergänzen, complete v.	86
Ergänzungen pl., addenda pl.	10
Ergänzungs-, supplementary	332
Erinnerungen pl., memoirs	197
erläutern, comment[1] v.; explain	81; 140
Erläuterung, commentary; explanation	82; 141
erscheinen, be published	272
erscheint nicht mehr, ceased publication	61
Erscheinungsvermerk, imprint	163
Erstausgabe, original edition	227
erste, first	147
erweitern, enlarge	133
erzählende Dichtung, fiction	146
Erzählung, short story; story	309; 324
Essay, essay[2]	136
Exemplar, copy[2]	97
Fabel, fable	143
Fachbibliographie, special bibliography	317
Fachbibliothek, special library	318
Faksimile, facsimile	144
Fälschung, forgery	151
Fassung, version	364
Festschrift, memorial volume	199
Figur, picture	241
Flugblatt, flysheet	148
fortsetzen, continue	92
Fortsetzung, sequel	304
Fragment, fragment	152
Freiexemplar, deposit copy	112
für, for	150
Fußnote, footnote	149
Gebrauch, use n.	363
gebrauchen, use v.	363
Gedächtnisrede, memorial speech	198
geheftetes Buch, paperback	232
gesammelte Werke pl., complete works pl.	87
Gesamtausgabe, complete works pl.	87
Gesamtkatalog, union catalogue	358
Gesang, song	314
Geschichte, history	157
Gesellschaft, society	313

Gesetz, law	177
gesetzliche Pflichtablieferung, copyright deposit	99
gewidmet, dedicated	108
Glossar, glossary	154
gravieren, engrave	131
groß, large	176
auf Grund des ..., based/founded on the ...	41
Handbuch, handbook	156
Handbücherei, reference library	279
Handschrift, manuscript	192
Haupt-, chief	66
Hauptzettel, main card	191
Heft, booklet; number[2]	50; 215
heften, sew	306
Heldendichtung, epic	134
Heldengedicht, epic	134
herausgeben, edit[2]	123
Herausgeber, editor[2]	126
Hirtengedicht, idyll	159
Hochzeitslied, bridal/nuptial song	53
Holzschnitt, woodcut	370
Holztafeldruck, woodcut	370
Hymne, anthem	24
Idylle, idyll	159
Ikonographie, iconography	158
illustrieren, illustrate	160
Imitation, imitation	161
Impressum, imprint	163
in, in	164
Inhalt, contents pl.	91
Inhaltsverzeichnis, table of contents	334
Initiale, initial	168
Inkunabel, incunabula	166
Inserat, advertisement	11
Institut, institute	170
Interimsaufnahme, temporary card	336
Interimszettel, temporary card	336
Inventur, inventory	173
Jahr, year	376
Jahrbuch, year-book	377
Jahrbücher pl., annals pl.	19
Jahrgang, volume[2]	367
Jahrhundert, century	62

APPENDIX

CALENDAR UNITS — TEMPORA

Januar, Jänner, January	I	**Dezember,** December	XII	**halbjährlich,** half-yearly,	
Februar, Feber, February	II	**Montag,** Monday	XIII	semi-annual	XXII
März, March	III	**Dienstag,** Tuesday	XIV	**vierteljährlich,** quarterly	XXIII
April, April	IV	**Mittwoch,** Wednesday	XV	**Monat,** month	XXIV
Mai, May	V	**Donnerstag,** Thursday	XVI	**monatlich,** monthly	XXV
Juni, June	VI	**Freitag,** Friday	XVII	**halbmonatlich,** fortnightly,	
Juli, July	VII	**Samstag, Sonnabend,**		semi-monthly	XXVI
August, August	VIII	Saturday	XVIII	**Woche,** week	XXVII
September, September	IX	**Sonntag,** Sunday	XIX	**wöchentlich,** weekly	XXVIII
Oktober, October	X	**Jahr,** year	XX	**Tag,** day	XXIX
November, November	XI	**jährlich,** annual, yearly	XXI	**täglich,** daily	XXX

NUMERALS — NUMERI

1 **eins**	one	16 **sechzehn**	sixteen
2 **zwei**	two	17 **siebzehn**	seventeen
3 **drei**	three	18 **achtzehn**	eighteen
4 **vier**	four	19 **neunzehn**	nineteen
5 **fünf**	five	20 **zwanzig**	twenty
6 **sechs**	six	21 **einundzwanzig**	twenty-one
7 **sieben**	seven	30 **dreißig**	thirty
8 **acht**	eight	40 **vierzig**	forty
9 **neun**	nine	50 **fünfzig**	fifty
10 **zehn**	ten	60 **sechzig**	sixty
11 **elf**	eleven	70 **siebzig**	seventy
12 **zwölf**	twelve	80 **achtzig**	eighty
13 **dreizehn**	thirteen	90 **neunzig**	ninety
14 **vierzehn**	fourteen	100 **hundert**	hundred
15 **fünfzehn**	fifteen	200 **zweihundert**	two hundred

300 dreihundert	three hundred	16. sechzehnte	sixteenth
400 vierhundert	four hundred	17. siebzehnte	seventeenth
500 fünfhundert	five hundred	18. achtzehnte	eighteenth
600 sechshundert	six hundred	19. neunzehnte	nineteenth
700 siebenhundert	seven hundred	20. zwanzigste	twentieth
800 achthundert	eight hundred	21. einundzwanzigste	twenty-first
900 neunhundert	nine hundred	30. dreißigste	thirtieth
1000 tausend	thousand	40. vierzigste	fortieth
1. erste	first	50. fünfzigste	fiftieth
2. zweite	second	60. sechzigste	sixtieth
3. dritte	third	70. siebzigste	seventieth
4. vierte	fourth	80. achtzigste	eightieth
5. fünfte	fifth	90. neunzigste	ninetieth
6. sechste	sixth	100. hundertste	hundredth
7. siebente	seventh	200. zweihundertste	two hundredth
8. achte	eighth	300. dreihundertste	three hundredth
9. neunte	ninth	400. vierhundertste	four hundredth
10. zehnte	tenth	500. fünfhundertste	five hundredth
11. elfte	eleventh	600. sechshundertste	six hundredth
12. zwölfte	twelfth	700. siebenhundertste	seven hundredth
13. dreizehnte	thirteenth	800. achthundertste	eight hundredth
14. vierzehnte	fourteenth	900. neunhundertste	nine hundredth
15. fünfzehnte	fifteenth	1000. tausendste	thousandth

GERMAN ALPHABET

LITTERAE GERMANICAE

A a, Ä ä	O o, Ö ö
B b	P p
C c	Q q
D d	R r
E e	S s, ß
F f	T t
G g	U u, Ü ü
H h	V v
I i	W w
J j	X x
K k	Y y
L l	Z z
M m	
N n	

RUSSIAN — RUSSICA

desjatičnaja klassifikacija (десятичная классифи-
кация), decimal classification   107

detskaja kniga (детская книга), children's book   67

dezideraty (дезидераты) *pl.*, desiderata *pl.*   113

dialog (диалог), dialog   114

dissertacija (диссертация), thesis   340

dlja (для), for   150

dnevnik (дневник), diary   115

dobavlenie (добавление), addenda   10

dobavljať (добавлять), complete   86

dobavočnoe opisanie (добавочное описание), added
entry   9

doklad (доклад), lecture; report[1]   179; 283

dokument (документ), document   117

dokumentacija (документация), documentation   118

dopolnenie (дополнение), addenda; supplement 10, 331

dopolnitelnyj (дополнительный), supplementary   332

dopolnjať (дополнять), complete *v.*   86

drama (драма), play   246

drevnij (древний), ancient   17

dublet (дублет), duplicate   121

dvujazyčnyj (двуязычный), bilingual   44

edinstvennyj (единственный), only   223

ėkzempljar (экземпляр), copy[2]   97

ėlegia (элегия), elegy   129

ėnciklopedija (энциклопедия), encyclopaedia   130

ėpičeskaja poėma (эпическая поэма), epic   134

ėpilog (эпилог), epilogue   135

ėpitalama (эпиталама), bridal/nuptial song   53

ėpopeja (эпопея), epic   134

ėpos (эпос) epic   134

ėskiz (эскиз), sketch   312

ėsse (эссе), essay[1]   136

ėstamp (эстамп), copperplate engraving   95

ėtjud (этюд), study   326

ežednevnik (ежедневник), newspaper   209

ežegodnik (ежегодник), annals *pl.*; year-book 19; 377

faksimile (факсимиле), facsimile   144

falsifikacija (фальсификация), forgery   151

familija (фамилия), name   206

foto (фото), photograph   240

foto(grafija) (фотография), photograph   240

fotosnimok (фотоснимок), photograph   240

fototipija (фототипия), collotype   76

fragment (фрагмент), fragment   152

gazeta (газета), newspaper   209

gimn (гимн), anthem   24

glava (глава), chapter   65

glavnyj (главный), chief   66

glossarij (глоссарий), glossary   154

glubokaja pečať s mednych form (глубокая печать
с медных форм), copperplate   94

god (год), year   376

godovoj komplekt (годовой комплект), volume[2]   367

god pečatanija (год печатания), date of printing   106

gorodskaja biblioteka (городская библиотека), city
library   68

gosudarstvennaja biblioteka (государственная библиотека), state library   322

grafa (графа), column   77

grammofonnaja plastinka (граммофонная плас-
тинка), record   276

gravirovať (гравировать), engrave   131

gravjura (гравюра), engraving   132

gravjura na dereve (гравюра на дереве), woodcut 370

gravjura na medi (гравюра на меди), copperplate
engraving; etching   95; 137

i (и), and   18

idillija (идиллия), idyll   159

ikonografija (иконография), iconography   158

illjustracija (иллюстрация), picture   241

illjustrirovať (иллюстрировать), illustrate; represent   160; 284

imitacija (имитация), imitation   161

imja (имя), name   206

indeks (индекс), index; location   167; 189

indeksacija (индексация), notation   211

inicial (инициал), initial   168

inkunabula (инкунабула), incunabula *pl.*   166

institut (институт), institute   170

inventar' (инвентарь), inventory   173

iskusstvo (искусство), art   30

ispolzovať (использовать), use *v.*   363

ispravljať (исправлять), correct   101

issledovanie, (исследование) study   326

istočnik (источник), source   316

istorija (история), history   157

iz (из), of   218

izbrannoe (избранное), selection   303

izdanie (издание), edition   124

APPENDIX

CALENDAR UNITS — TEMPORA

janvar' (январь), January	I
fevral' (февраль), February	II
mart (март), March	III
aprel' (апрель), April	IV
maj (май), May	V
ijun' (июнь), June	VI
ijul' (июль), July	VII
avgust (август), August	VIII
sentjabr' (сентябрь), September	IX
oktjabr' (октябрь), October	X
nojabr' (ноябрь), November	XI
dekabr' (декабрь), December	XII
ponedeľnik (понедельник), Monday	XIII
vtornik (вторник), Tuesday	XIV
sreda (среда) Wednesday	XV
četverg (четверг), Thursday	XVI
pjatnica (пятница), Friday	XVII
subbota (суббота), Saturday	XVIII
voskresen'e (воскресенье), Sunday	XIX
god (год), year	XX
ežegodnyj (ежегодный), godovoj (годовой), annual, yearly	XXI
polugodovoj (полугодовой), šestimesjačnyj (шестимесячный), semi-annual	XXII
trëhmesjačnyj (трёхмесячный), quarterly	XXIII
mesjac (месяц), month	XXIV
ežemesjačnyj (ежемесячный), monthly	XXV
polumesjačnyj (полумесячный), fortnightly, semi-monthly	XXVI
nedelja (неделя), week	XXVII
eženedeľnyj (еженедельный), weekly	XXVIII
den' (день), day	XXIX
ežednevnyj (ежедневный), daily	XXX

NUMERALS — NUMERI

1	odin, odna, odno (один, одна, одно)	one
2	dva, dve (два, две)	two
3	tri (три)	three
4	četyre (четыре)	four
5	pjať (пять)	five
6	šesť (шесть)	six
7	sem' (семь)	seven
8	vosem' (восемь)	eight
9	devjať (девять)	nine
10	desjať (десять)	ten
11	odinnadcať (одиннадцать)	eleven
12	dvenadcať (двенадцать)	twelve
13	trinadcať (тринадцать)	thirteen

218

14	četyrnadcaſ (четырнадцать)	fourteen
15	pjatnadcaſ (пятнадцать)	fifteen
16	šestnadcaſ (шестнадцать)	sixteen
17	semnadcaſ (семнадцать)	seventeen
18	vosemnadcaſ (восемнадцать)	eighteen
19	devjatnadcaſ (девятнадцать)	nineteen
20	dvadcaſ (двадцать)	twenty
21	dvadcaſ odin (двадцать один)	twenty-one
30	tridcaſ (тридцать)	thirty
40	sorok (сорок)	forty
50	pjaſdesjat (пятьдесят)	fifty
60	šesſdesjat (шестьдесят)	sixty
70	sem'desjat (семьдесят)	seventy
80	vosem'desjat (восемьдесят)	eighty
90	devjanosto (девяносто)	ninety
100	sto (сто)	hundred
200	dvesti (двести)	two hundred
300	trista (триста)	three hundred
400	četyresta (четыреста)	four hundred
500	pjaſsot (пятьсот)	five hundred
600	šesſsot (шестьсот)	six hundred
700	sem'sot (семьсот)	seven hundred
800	vosem'sot (восемьсот)	eight hundred
900	devjaſsot (девятьсот)	nine hundred
1000	tysjača (тысяча)	thousand
1.	pervyj, -aja, -oe (первый, -ая, -ое)	first
2.	vtoroj, -aja, -oe (второй, -ая, -ое)	second
3.	tretij (третий)	third
4.	četvěrtyj (четвёртый)	fourth
5.	pjatyj (пятый)	fifth
6.	šestoj (шестой)	sixth
7.	sed'moj (седьмой)	seventh
8.	vos'moj (восьмой)	eighth
9.	devjatyj (девятый)	ninth
10.	desjatyj (десятый)	tenth
11.	odinnadcatyj (одиннадцатый)	eleventh
12.	dvenadcatyj (двенадцатый)	twelfth
13.	trinadcatyj (тринадцатый)	thirteenth
14.	četyrnadcatyj (четырнадцатый)	fourteenth
15.	pjatnadcatyj (пятнадцатый)	fifteenth
16.	šestnadcatyj (шестнадцатый)	sixteenth
17.	semnadcatyj (семнадцатый)	seventeenth
18.	vosemnadcatyj (восемнадцатый)	eighteenth
19.	devjatnadcatyj (девятнадцатый)	nineteenth
20.	dvadcatyj (двадцатый)	twentieth
21.	dvadcaſ pervyj (двадцать первый)	twenty-first
30.	tridcatyj (тридцатый)	thirtieth
40.	sorokovoj (сороковой)	fortieth
50.	pjatidesjatyj (пятидесятый)	sixtieth
60.	šestidesjatyj (шестидесятый)	fiftieth
70.	semidesjatyj (семидесятый)	seventieth
80.	vos'midesjatyj (восьмидесятый)	eightieth
90.	devjanostyj (девяностый)	ninetieth
100.	sotyj (сотый)	hundredth
200.	dvuhsotyj (двухсотый)	two hundredth
300.	trěhsotyj (трёхсотый)	three hundredth
400.	četyrěhsotyj (четырёхсотый)	four hundredth
500.	pjatisotyj (пятисотый)	five hundredth
600.	šestisotyj (шестисотый)	six hundredth
700.	semisotyj (семисотый)	seven hundredth
800.	vos'misotyj (восьмисотый)	eight hundredth
900.	devjatisotyj (девятисотый)	nine hundredth
1000.	tysjačnyj (тысячный)	thousandth

RUSSIAN (CYRILLIC) ALPHABET. TRANSLITERATION
LITTERAE RUSSICAE (CYRILLIANAE). TRANSLITTERATIO

А а,	a	К к,	k	Х х,	~~h, ch~~ *KH*		
Б б,	b	Л л,	l	Ц ц,	c *ts*		
В в,	v	М м,	m	Ч ч,	č *CH*		
Г г,	g	Н н,	n	Ш ш,	š *SH*		
Д д,	d	О о,	o	Щ щ,	šč *SHCH*		
Е е, Ё ё	e, ё	П п,	p	~~Ъ ъ,~~	"		
Ж ж,	ž *ZH*	Р р,	r	Ы ы	y		
З з,	z	С с,	s	Ь ь,	'		
И и,	i	Т т,	t	Э э,	ė		
Й й,	j	У у,	u	Ю ю,	ju		
		Ф ф,	f	Я я,	ja		

SPANISH – HISPANICA

SPANISH

disco, record[1]		276
discurso, speech		319
discurso conmemorativo, memorial speech		198
diseño, drawing; sketch		120; 312
disertación, thesis; treatise		340; 352
disertar, treat		351
en **diversos sitios,** passim		237
documentación, documentation		118
documento, document		117
drama, play		246
dudoso, doubtful		119
duplicado, copy[1]; duplicate		96; 121
edición, edition		124
edición original, original edition		227
edición príncipe, original edition		227
editar, edit[1]; edit[2]; publish		122; 123; 271
editor, editor[2]; publisher[1]		126; 273
editorial, publishing house		274
ejemplar, copy[2]		97
ejemplar de depósito legal, deposit copy		112
ejemplar mecanografiado, typescript		353
el, the		339
elaborar, elaborate		128
elegía, elegy		129
elegir, select		302
elogio, panegyric		230
emplear, use *v.*		363
empleo, use *n.*		362
en, in		164
encabezamiento secundario (España), added entry		9
enciclopedia, encyclopaedia		130
encuadernación, binding		47
encuadernar, bind		45
encuadernar en rústica, sew		306
enmendar, correct		101
ensayo, essay[1]		136
entero, complete *a.*		85
entrada secundaria (Hispanoamérica), added entry		9
epílogo, epilogue		135
epístola, letter		181
epitalamio, bridal/nuptial song		53
epítome, extract		142
epopeya, epic		134
error de imprenta, printer's error		265
error tipográfico, printer's error		265

esbozo, sketch		312
escala, scale		293
escoger, select		302
escribir, write		372
escritor, writer		374
escritura, script		299
escritura Braille, braille printing		52
escuela, school		295
estampa, copperplate engraving		95
estampar, print[1]		261
estandartización, standardisation		321
estatuto, law; statute[1]		177; 323
estenografía, shorthand		308
estrofa, strophe		325
estudio, study		326
exhibición, exhibition		139
explicación, explanation		141
explicar, explain		140
exposición, exhibition		139
extracto, extract		142
fábula, fable		143
facsímile, facsimile		144
falsificación, forgery		151
fascículo, number[2]		215
fecha de impresión, date of printing		106
ficha, card		56
ficha de referencia, reference card		278
ficha principal, main card		191
ficha provisional, temporary card		336
ficha secundaria, added entry		9
figura, picture		241
figurar, represent		284
folleto, flysheet		148
fotografía, photograph		240
fototipia, collotype		76
fragmento, fragment		152
fuente, source		316
fuerza del cuerpo, type size		354
gaceta, newspaper		209
general, chief		66
glosario, glossary		154
grabado, engraving		132
grabado al agua fuerte, etching		137

resumen, summary		330
resumir, summarize		329
retrato, portrait		253
reunir, summarize		329
revisar, revise[1]		289
revisión, revision		290
revista, periodical[2]; review		239; 287
revista de modas, fashion magazine		145

salir, *be* published		272
sección, section		301
selección, selection		303
sentencia, adage		7
serie, series		305
servicio de abono, subscription		328
servicio de canje, exchange centre		138
sesión, sitting		311
seudónimo, pseudonym		269
siglo, century		62
signatura, signature		310
signatura topográfica, location mark		189
sin, without		369
s. l. s. a. → *sin* lugar sin año		
sobre, on		222
sociedad, society		313
subscripción, subsription		328
sumario, table of contents		334
suplementario, supplementary		332
suplemento, addenda *pl.;* supplement		10; 331

tabla, table		333
tabla de materias, table of contents		334
taller de encuadernación, bindery		46
taller de encuadernador, bindery		46
tebeo, comic (news)paper		79
texto, text		337
tipografía, printing office; typography		226; 355
tirada, impression		162

tirada aparte, offprint		221
tiraje, impression		162
tiraje aparte, offprint		221
título, title		341
título actual, running title		292
tomo, volume[1]		366
traducción, translation		349
traducir, translate		348
tragedia, tragedy		345
trama, screen		298
transcripción, transcription		347
transliteración, transliteration		350
tratado, treatise		352
tratar, treat		351
trozo, fragment		152

un, a, an		1
una, a, an		1
único, only		223
universidad, university		359
uno, a, an		1
usar, use *v.*		363
uso, use *n.*		362

variante, version		364
versión, version		364
verso[1], poem		248
verso[2], verso		365
viejo, ancient		17
vocabulario, dictionary; glossary		116; 154
volumen, volume[1]		366
volumen conmemorativo, memorial volume		199

xerografía, xerography		375
xilografía, woodcut		370

y, and		18

APPENDIX

CALENDAR UNITS — TEMPORA

enero, January	I	diciembre, December	XII	semestral, half-yearly,
febrero, February	II	lunes, Monday	XIII	semi-annual	XXII
marzo, March	III	martes, Tuesday	XIV	trimestral, quarterly	XXIII
abril, April	IV	miércoles, Wednesday	XV	mes, month	XXIV
mayo, May	V	jueves, Thursday	XVI	mensual, monthly	XXV
junio, June	VI	viernes, Friday	XVII	semimensual, fortnightly,
julio, July	VII	sábado, Saturday	XVIII	semi-monthly	XXVI
agosto, August	VIII	domingo, Sunday	XIX	semana, week	XXVII
septiembre, September	IX	año, year	XX	semanal, weekly	XXVIII
octubre, October	X	anual, annual, yearly	XXI	día, day	XXIX
noviembre, November	XI			cotidiano, diario, daily	XXX

NUMERALS — NUMERI

1	uno, una, un	one		16	diez y seis	sixteen
2	dos	two		17	diez y siete	seventeen
3	tres	three		18	diez y ocho	eighteen
4	cuatro	four		19	diez y nueve	nineteen
5	cinco	five		20	veinte	twenty
6	seis	six		21	veinte y uno	twenty-one
7	siete	seven		30	treinta	thirty
8	ocho	eight		40	cuarenta	forty
9	nueve	nine		50	cincuenta	fifty
10	diez	ten		60	sesenta	sixty
11	once	eleven		70	setenta	seventy
12	doce	twelve		80	ochenta	eighty
13	trece	thirteen		90	noventa	ninety
14	catorce	fourteen		100	cien(to)	hundred
15	quince	fifteen		200	doscientos	two hundred

300	trescientos	three hundred	16.	décimosexto	sixteenth
400	cuatrocientos	four hundred	17.	décimoséptimo	seventeenth
500	quinientos	five hundred	18.	décimooctavo	eighteenth
600	seiscientos	six hundred	19.	décimonono	nineteenth
700	setecientos	seven hundred	20.	vigésimo	twentieth
800	ochocientos	eight hundred	21.	vigésimo primo	twenty-first
900	novecientos	nine hundred	30.	trigésimo	thirtieth
1000	mil	thousand	40.	cuadragésimo	fortieth
1.	primer(o), -ra	first	50.	quincuagésimo	fiftieth
2.	segundo, -da	second	60.	sexagésimo	sixtieth
3.	tercer(o), tercio	third	70.	septuagésimo	seventieth
4.	cuarto	fourth	80.	octogésimo	eightieth
5.	quinto	fifth	90.	nonagésimo	ninetieth
6.	sexto	sixth	100.	centésimo	hundredth
7.	séptimo	seventh	200.	ducentésimo	two hundredth
8.	octavo	eighth	300.	tricentésimo	three hundredth
9.	noveno, nono	ninth	400.	cuadringentésimo	four hundredth
10.	décimo	tenth	500.	quingentésimo	five hundredth
11.	undécimo	eleventh	600.	sexcentésimo	six hundredth
12.	duodécimo	twelfth	700.	septingentésimo	seven hundredth
13.	décimotercio	thirteenth	800.	octingentésimo	eight hundredth
14.	décimocuarto	fourteenth	900.	noningentésimo	nine hundredth
15.	décimoquinto	fifteenth	1000.	milésimo	thousandth

SPANISH ALPHABET

LITTERAE HISPANICAE

A a	J j	R r
B b	K k	S s
C c	L l	T t
Ch ch	Ll ll	U u
D d	M m	V v
E e	N n	X x
F f	Ñ ñ	Y y
G g	O o	Z z
H h	P p	
I i	Q q	

BULGARIAN — BULGARICA

inventar (инвентар), inventory 173
istinski (истински), authentic 34
istorija (история), history 157
izbiram (избирам), select 302
izbrani otkâsi (избрани откъси), selection 303
izdanie (издание), edition 124
izdanie s meka podvârzija (издание с мека подвързия), paperback 232
izdatel (издател), editor; publisher 126; 273
izdatelski danni (издателски данни) pl. imprint 163
izdavam (издавам), edit; publish 123; 271
izdavam se (издавам се), be published 272
izkustvo (изкуство), art 30
izlizam (излизам), be published 272
izložba (изложба), exhibition 139
izobrazjavam (изобразявам), represent 284
izrabotvam (изработвам), elaborate 128
iztočnik (източник), source 316
izvadka (извадка), extract 142
izvestija (известия) pl., bulletin; transactions pl. 54; 346
izvlečenie (извлечение), extract 142
izvor (извор), source 316

-ja (-я), the 339
-jat (-ят), the 339
jubileen sbornik (юбилеен сборник), memorial
   volume 199

kartička (картичка), card 56
kartina (картина), picture 241
kartonče (картонче), card 56
katalog (каталог), catalog(ue) n. 58
katalogizacija (каталогизация), cataloging 60
kataloghiram (каталогизирам), catalog(ue) v. 59
kataložno kartonče (каталожно картонче), card 56
klasifikacija (класификация), classification 69
klauza (клауза), clouse 70
kniga (книга), book 49
knigohranilište (книгохранилище), stack-room 320
knigoizdatelstvo (книгоиздателство), publishing
   house 274
knigoobmenna služba (книгообменна служба),
   exchange centre 138
knigoveznica (книговезница), bindery 46
knigozaemane (книгозаемане), lending 180
knigoznanie (книгознание), bibliology 43

knižarnica (книжарница), bookshop 51
knižka (книжка), booklet; number[2] 50; 215
knižnina (книжнина), literature 186
kodeks (кодекс), codex 71
kola (кола), sheet 307
kolektiven avtor (колективен автор), corporate
   author 100
kolona (колона), column 77
kolontitul (колонтитул), running title 292
komedija (комедия), comedy 78
komentar (коментар), commentary 82
komentiram (коментирам), comment 81
komiks (комикс), comic strip 80
komisija (комисия), committee 83
komitet (комитет), committee 83
komplekten (комплектен), complete a. 85
komplektuvam (комплектувам), complete v. 86
konferencija (конференция), conference 88
kongres (конгрес), congress 89
korekcija (корекция), correction 102
korespondencija (кореспонденция), correspondence 103
kritika (критика), critique 104
kritikuvam (критикувам), criticize 105
kserografija (ксерография), xerography 375
kuplet (куплет), strophe 325

lekcija (лекция), lecture 179
na lenta (на лента), on tape 335
letopisi (летописи) pl., annals pl. 19
lice na list (лице на лист), recto 277
list (лист), leaf; sheet 178; 307
listovo izdanie (листово издание), loose-leaf book 190
literatura (литература), literature 186
literaturno nasledstvo (литературно наследство),
   literary remains pl. 185
litografija (литография), litograph(y) 187

maksima (максима), adage 7
malâk (малък), little 188
masova biblioteka (масова библиотека), public
   library 270
mašinopisen ekzempljar (машинописен екземпляр),
   typescript 353
maštab (мащаб), scale 293
medna gravjura (медна гравюра), copperplate 94

memoari (мемоари) *pl.*, memoirs *pl.* 197
memorandum (меморандум), memoir 196
menja se (меня се), change² 64
mikrofilm (микрофилм), microfilm 200
miniatjura (миниатюра), miniature 201
mjasto (място), place 243
bez mjasto i godina, b. m. i g. (без място и година, б. м. и г.) no place no date 210
mnogoezičen (многоезичен), polyglot 252
moden žurnal (моден журнал), fashion magazine 145
modno spisanie (модно списание), fashion magazine 145
monografija (монография), monograph(y) 204
muzika (музика), music 205

na (на), in; of; on; to 164; 218; 222; 344
nabor (набор), matter 195
nacionalna biblioteka (национална библиотека), national library 207
narâčnik (наръчник), handbook 156
naredba (наредба), order¹ 224
nareždam po azbučen red (нареждам по азбучен ред), alphabetize 16
narodna biblioteka (народна библиотека), national library 207
naučno-fantastična literatura (научно-фантастична литература), science fiction 297
nauka (наука), science 296
neizdaden (неиздаден), unpublished 361
neizmenen (неизменен), unchanged 356
nekomplekten (некомплектен), incomplete 165
nekrolog (некролог), obituary notice 216
nepâlen (непълен), incomplete 165
nesmenen (несменен), unchanged 356
nomer (номер), number¹ 214
normalizacija (нормализация), standardisation 321
notni izdanija (нотни издания) *pl.*, printed music 263
nov (нов), new 208
novela (новела), short story 309
novo izdanie (ново издание), reprint 285
novo stereotipno izdanie (ново стереотипно издание), reprint 285

objasnenie (обяснение), explanation 141
objasnjavam (обяснявам), explain 140
objavlenie (обявление), advertisement 11
obratna strana (обратна страна), verso 365

obraz (образ), picture 241
obštoobrazovatelna biblioteka (общообразователна библиотека), public library 270
očerk (очерк), sketch 312
oficialen (официален), official 220
opisanie na knigi (описание на книги), cataloging 60
opisanie pod zaglavie (описание под заглавие), title entry 342
oproverženie (опровержение), refutation 281
originalen (оригинален), original 226
vâz osnova na (въз основа на), based/founded on the . . . 41
osnoven fiš (основен фиш), main card 191
osnovna kartička (основна картичка), main card 191
ot (от), by; from; of 55; 153; 218
otčet (отчет), report 283
otdel (отдел), section 301
otdelenie (отделение), section 301
otdelen otpečatâk (отделен отпечатък), offprint 221
otkâslek (откъслек), fragment 152
otnosno (относно), about 3
otpečatâk (отпечатък), impression 162
otpečatvam (отпечатвам), print 261
otraslova bibliografija (отраслова библиография), special bibliography 317
otraslova biblioteka (отраслова библиотека), special library 318
otziv (отзив), critique 104

paginacija (пагинация), paging 229
papirus (папирус), papyrus 233
paragraf (параграф), paragraph 234
parče (парче), piece 242
pâlen (пълен), complete *a.* 84
pâlno izdanie (пълно издание), complete works 87
pâlno sâbranie na sâčinenijata (пълно събрание на съчиненията), complete works 87
pârvi (първи), first 147
pârvo izdanie (първо издание), original edition 227
pârvopečatna kniga (първопечатна книга), incunabula *pl.* 166
pâtevoditel (пътеводител), guide(-book) 155
pečat (печат), press¹ 260
pečatam (печатам), print¹ 261
pečatar (печатар), printer 264
pečatna greška (печатна грешка), printer's error 265
pečatnica (печатница), printing office 266

pečatno (proizvedenie) (печатно произведение), printed matter 262

pergament (пергамент), parchment 235

periodičen (периодичен), periodical[1] 238

personalna bibliografija (персонална библиогра-фия), author bibliography 36

pesen (песен), song 314

pesnopojka (песнопойка), songbook 315

piesa (пиеса), play 246

pisane (писане), script 299

pisatel (писател), writer 374

pismo (писмо), letter 181

piša (пиша), write 372

piša stihove (пиша стихове), write poetry 373

plagiat (плагиат), plagiarism 244

po (по), about; after; on 3; 12; 222

pod (под), under 357

podgotvjam (подготвям), prepare 257

podpiska (подписка), subscription 328

podpravka (подправка), forgery 151

podrâčna sbirka (подръчна сбирка), reference library 279

podreždam (подреждам), arrange 29

podšivam (подшивам), sew 306

podvârzija (подвързия), binding 47

podvârzvam (подвързвам), bind 45

poet (поет), poet 249

poezija (поезия), poetry 250

pogovorka (поговорка), adage 7

pohvalno slovo (похвално слово), panegyric 230

polemika (полемика), polemic 251

pole na kniga (поле на книга), margin 194

poliglot (полиглот), polyglot 252

poligrafija (полиграфия), typography 355

polzuvane (ползуване), use n. 362

popravjam (поправям), correct 101

popravka (поправка), correction 102

porâčka (поръчка), order[2] 225

poredica (поредица), series 305

portret (портрет), portrait 253

poslepis (послепис), postscript 255

posleslovie (послесловие), epilogue 135

posmârten (посмъртен), posthumous 254

posmârtni proizvedenija (посмъртни произведения) pl., literary remains pl. 185

postanovlenie (постановление), order[1] 224

posveštenie (посвещение), dedication 109

posveten (посветен), dedicated 108

poziv (позив), flysheet 148

pravilnik (правилник), statute 323

predgovor (предговор), preface 256

predmetna rubrika (предметна рубрика), subject heading 327

predpisanie (предписание), prescription 258

predsedatel (председател), president 259

pregled (преглед), review n.; revision 287; 290

pregleždam (преглеждам), revise 289

preizdavane, (преиздаване), reprint 285

prepečatvane (препечатване), reprint 285

prepis (препис), copy[1] 96

prepiska (преписка), correspondence 103

prepratka (препратка), reference card 278

prerabotvam (преработвам), rewrite 291

presa (преса) press[1] 260

preveždam (превеждам), translate 348

prevod (превод), translation 349

pribavjam (прибавям), annex 20

pribavka (прибавка), appendix 26

prigotvjam (приготвям), prepare 257

prigotvjam za pečat (приготвям за печат), edit 122

prikazka (приказка), fable 143

priložen (приложен), applied 27

priloženie (приложение), appendix; supplement 26; 331

prinos (принос), contribution 93

pripiska (приписка), postscript 255

priturka (притурка), appendix; supplement 26; 331

priveždam v red (привеждам в ред), arrange 29

prodâlgovat (продълговат), oblong 217

prodâlžavam (продължавам), continue 92

prodâlženie (продължение), sequel 304

proizvedenie (произведение), work 371

promenjam (променям), change[1] 63

promenjam se (променям се), change[2] 64

protokol (протокол), minutes pl. 202

proučvane (проучване), study 326

proza (проза), prose 267

psevdonim (псевдоним), pseudonym 269

raster (растер), screen 298

razgleždam (разглеждам), criticize; review v.; treat 105; 288; 351

razjasnjavam (разяснявам), comment; explain 81; 140

razkaz (разказ), short story; story 309; 324

razmnožavane (размножаване), reproduction 286

razni (разни) pl., miscellanea pl. 203

razrabotvam (разработвам), elaborate 128

APPENDIX

CALENDAR UNITS — TEMPORA

januari (януари), January	I	**noemvri** (ноември), November	XI	**godišen** (годишен), yearly	XXI
fevruari (февруари), February	II	**dekemvri** (декември), December	XII	**polugodišen** (полугодишен), half-yearly	XXII
mart (март), March	III	**ponedelnik** (понеделник), Monday	XIII	**trimesečen** (тримесечен), quarterly	XXIII
april (април), April	IV	**vtornik** (вторник), Tuesday	XIV	**mesec** (месец), month	XXIV
maj (май), May	V	**srjada** (сряда), Wednesday	XV	**mesečen** (месечен), monthly	XXV
juni (юни), June	VI	**četvârtâk** (четвъртък), Thursday	XVI	**polumesečen** (полумесечен), fortnightly	XXVI
juli (юли), July	VII	**petâk** (петък), Friday	XVII	**sedmica** (седмица), week	XXVII
avgust (август), August	VIII	**sâbota** (събота), Saturday	XVIII	**sedmičen** (седмичен), weekly	XXVIII
septemvri (септември), September	IX	**nedelja** (неделя), Sunday	XIX	**den** (ден), day	XXIX
oktomvri (октомври), October	X	**godina** (година), year	XX	**vsekidneven** (всекидневен), daily	XXX

NUMERALS — NUMERI

1 **edin, edna, edno** (един, една, едно)	one	9 **devet** (девет)	nine
2 **dva, dve** (два, две)	two	10 **deset** (десет)	ten
3 **tri** (три)	three	11 **edinadeset** (единадесет)	eleven
4 **četiri** (четири)	four	12 **dvanadeset** (дванадесет)	twelve
5 **pet** (пет)	five	13 **trinadeset** (тринадесет)	thirteen
6 **šest** (шест)	six	14 **četirinadeset** (четиринадесет)	fourteen
7 **sedem** (седем)	seven	15 **petnadeset** (петнадесет)	fifteen
8 **osem** (осем)	eight	16 **šestnadeset** (шестнадесет)	sixteen
		17 **sedemnadeset** (седемнадесет)	seventeen

18 osemnadeset (осемнадесет)	eighteen	10. deseti (десети)	tenth
19 devetnadeset (деветнадесет)	nineteen	11. edinadeseti (единадесети)	eleventh
20 dvadeset (двадесет)	twenty	12. dvanadeseti (дванадесети)	twelfth
21 dvadeset i edin (двадесет и един)	twenty one	13. trinadeseti (тринадесети)	thirteenth
30 trideset (тридесет)	thirty	14. četirinadeseti (четиринадесети)	fourteenth
40 četirideset (четиридесет)	forty	15. petnadeseti (петнадесети)	fifteenth
50 petdeset (петдесет)	fifty	16. šestnadeseti (шестнадесети)	sixteenth
60 šestdeset (шестдесет)	sixty	17. sedemnadeseti (седемнадесети)	seventeenth
70 sedemdeset (седемдесет)	seventy	18. osemnadeseti (осемнадесети)	eighteenth
80 osemdeset (осемдесет)	eighty	19. devetnadeseti (деветнадесети)	nineteenth
90 devetdeset (деветдесет)	ninety	20. dvadeseti (двадесети)	twentieth
100 sto (сто)	hundred	21. dvadeset i pârvi (двадесет и първи)	twenty first
200 dvesta (двеста)	two hundred	30. trideseti (тридесети)	thirtieth
300 trista (триста)	three hundred	40. četirideseti (четиридесети)	fortieth
400 četiristotin (четиристотин)	four hundred	50. petdeseti (петдесети)	fiftieth
500 petstotin (петстотин)	five hundred	60. šestdeseti (шестдесети)	sixtieth
600 šeststotin (шестстотин)	six hundred	70. sedemdeseti (седемдесети)	seventieth
700 sedemstotin (седемстотин)	seven hundred	80. osemdeseti (осемдесети)	eightieth
800 osemstotitn (осемстотин)	eight hundred	90. devetdeseti (деветдесети)	ninetieth
900 devetstotin (деветстотин)	nine hundred	100. stoten (стотен)	hundredth
1000 hiljada (хиляда)	thousand	200. dvestoten (двестотен)	two hundredth
1. pârvi, -a, -o (първи, -а, -о)	first	300. tristoten (тристотен)	three hundredth
2. vtori (втори)	second	400. četiristoten (четиристотен)	four hundredth
3. treti (трети)	third	500. petstoten (петстотен)	five hundredth
4. četvârti (четвърти)	fourth	600. šeststoten (шестстотен)	six hundredth
5. peti (пети)	fifth	700. sedemstoten (седемстотен)	seven hundredth
6. šesti (шести)	sixth	800. osemstoten (осемстотен)	eight hundredth
7. sedmi (седми)	sevenrh	900. devetstoten (деветстотен)	nine hundredth
8. osmi (осми)	eighth	1000. hiljaden (хиляден)	thousand
9. deveti (девети)	ninth		

BULGARIAN (CYRILIC) ALPHABET. TRANSLITERATION
LITTERAE BULGARICAE (CYRILLIANAE). TRANSLITTERATIO

А а,	a	И и,	i	Р р,	r	Ш ш,	š
Б б,	b	Й й,	j	С с,	s	Щ щ,	št
В в,	v	К к,	k	Т т,	t	Ъ ъ,	â
Г г,	g	Л л,	l	У у,	u	Ь ь,	’
Д д,	d	М м,	m	Ф ф,	f	Ю ю,	ju
Е е,	e	Н н,	n	Х х,	h	Я я,	ja
Ж ж,	ž	О о,	o	Ц ц,	c		
З з,	z	П п,	p	Ч ч,	č		

CROATIAN – CROATICA

narudžbina, order[2]	225
nasljedstvo, literary remains *pl.*	185
naslov, title	341
naslovni list, title page	343
nastavak, sequel	304
nastavljati, continue	92
natpis, title	341
nauka, science	296
neizdan, unpublished	361
nekompletan, defective	110
nekrolog, obituary notice	216
nepotpun, incomplete	165
nepromjenjiv, unchanged	356
nepromjenljiv, unchanged	356
nevjerodostojan, doubtful, dubious	119
normiranje, standardisation	321
note *pl.*, printed music	263
nov, new	208
novela, short story	309
novine *pl.*, newspaper	209

o, about; on	3; 222
obavijest, advertisement	11
objašnjavati, explain	140
objašnjenje, commentary; explanation	82; 141
objavljivati, *be* published	272
obraditi, elaborate	128
obveza dostave tiskanih stvari, copyright deposit	99
obvezan primjerak, deposit copy	112
ocjenjivati, criticize	105
od, by; from; of	55; 153; 218
odabrati, select	302
odbor, committee	83
odjeljak, section	301
odlomak, extract; fragment	142; 152
oglas, advertisement	11
ogled, essay[1]; study	136; 326
opaska, annotation; note[1]	22; 212
opovrgnuće, refutation	281
originalan, original	226
na osnovu..., based/founded on the...	41
osobna bibliografija, author bibliography	36
ostavština, literary remains *pl.*	185
otisak, impression; printed matter	162; 262
ovdje-ondje, passim	237
ovlastiti, authorize	37
označavanje, notation	211

pa, and	18
paginacija, paging	229
papir, paper[1]	231
papirus, papyrus	233
paragraf, paragraph	234
parče, piece	242
pergament, parchment	235
periodičan, periodical[1]	238
pisac, writer	374
pisanje, script	299
pisati, write	372
pisati pesme, write poetry	373
pismo, letter	181
pjesma, poem; song	248; 314
pjesmarica, songbook	315
pjesnik, poet	249
pjesništvo, poetry	250
plagijat, plagiarism	244
po, about; on	3; 222
pod, under	357
podaci *pl.*, contribution	93
podražavanje, imitation	161
poezija, poetry	250
poglavlje, chapter	65
pogovor, epilogue	135
pohvalna pjesma, panegyric	230
polemičan spis, polemic	251
popis, list	184
popraviti, correct	101
popuniti, complete *v.*	86
portret, portrait	253
poseban otisak, offprint	221
poslije, after	12
posmrtni, posthumous	254
posmrtni govor, memorial speech	198
postskriptum, postscript	255
posudba, lending	180
posvećen, dedicated	108
posveta, dedication	109
potpun, complete *a.*	85
povijest, history	157
pravi, authentic	34
pravilo, prescription; statute[1]	258; 323
predavanje, lecture	179
predgovor, preface	256
predmetna oznaka, subject heading	327
predsjednik, president	259
predstavljati, represent	284

APPENDIX

CALENDAR UNITS — TEMPORA

siječanj, January	I	**prosinac,** December	XII	**polugodišnji,** half-yearly,
veljača, February	II	**ponedjeljak,** Monday	XIII	semi-annual — XXII
ožujak, March	III	**utornik,** Tuesday	XIV	**tromjesečni,** quarterly — XXIII
travanj, April	IV	**srijeda,** Wednesday	XV	**mjesec,** month — XXIV
svibanj, May	V	**četvrtak,** Thursday	XVI	**mjesečni,** monthly — XXV
lipanj, June	VI	**petak,** Friday	XVII	**polumjesečni,** fortnightly,
srpanj, July	VII	**subota,** Saturday	XVIII	semi-monthly — XXVI
kolovoz, August	VIII	**nedelja,** Sundap	XIX	**sedmica, tjedan,** week — XXVII
rujan, September	IX	**godina,** year	XX	**sedmični, tjedni,** weekly — XXVIII
listopad, October	X	**gidišnji,** annual, yearly	XXI	**dan,** day — XXIX
studeni, November	XI			**dnevni,** daily — XXX

NUMERALS — NUMERI

1 **jedan, -dna, -dno**	one	16 **šesnaest**	sixteen
2 **dva**	two	17 **sedamnaest**	seventeen
3 **tri**	three	18 **osamnaest**	eighteen
4 **četiri**	four	19 **devetnaest**	nineteen
5 **pet**	five	20 **dvadeset**	twenty
6 **šest**	six	21 **dvadeset i jedan**	twenty-one
7 **sedam**	seven	30 **trideset**	thirty
8 **osam**	eight	40 **četrdeset**	forty
9 **devet**	nine	50 **pedeset**	fifty
10 **deset**	ten	60 **šezdeset**	sixty
11 **jedanaest**	eleven	70 **sedamdeset**	seventy
12 **dvanaest**	twelve	80 **osamdeset**	eighty
13 **trinaest**	thirteen	90 **devedeset**	ninety
14 **četrnaest**	fourteen	100 **sto**	hundred
15 **petnaest**	fifteen	200 **dvesta, dve stotine**	two hundred

300 trista, tri stotine	three hundred	16. šesnaesti	sixteenth
400 četiristo, četiri stotine	four hundred	17. sedamnaesti	seventeenth
500 petsto, pet stotina	five hundred	18. osamnaesti	eighteenth
600 šeststo	six hundred	19. devetnaesti	nineteenth
700 sedamsto	seven hundred	20. dvadeseti	twentieth
800 osamsto	eight hundred	21. dvadeset i prvi	twenty-first
900 devetsto	nine hundred	30. trideseti	thirtieth
1000 tisuća	thousand	40. četrdeseti	fortieth
1. prvi, -a, -o	first	50. pedeseti	fiftieth
2. drugi	second	60. šezdeseti	sixtieth
3. treći	third	70. sedamdeseti	seventieth
4. četvrti	fourth	80. osamdeseti	eightieth
5. peti	fifth	90. devedeseti	ninetieth
6. šesti	sixth	100. stoti	hundredth
7. sedmi	seventh	200. dvestoti	two hundredth
8. osmi	eighth	300. tristoti	three hundredth
9. deveti	ninth	400. četiristoti	four hundredth
10. deseti	tenth	500. petstoti	five hundredth
11. jedanaesti	eleventh	600. šeststoti	six hundredth
12. dvanaesti	twelfth	700. sedamstoti	seven hundredth
13. trinaesti	thirteenth	800. osamstoti	eight hundredth
14. četrnaesti	fourteenth	900. devetstoti	nine hundredth
15. petnaesti	fifteenth	1000. tisući	thousandth

CROATIAN ALPHABET

LITTERAE CROATICAE

A a	G g	O o
B b	H h	P p
C c	I i	R r
Č č	J j	S s
Ć ć	K k	Š š
D d	L l	T t
Dž dž	Lj lj	U u
Đ đ	M m	V v
E e	N n	Z z
F f	Nj nj	Ž ž

CZECH — BOHEMICA

konference, conference	88
kongres, congress	89
kopie, copy[1]	96
korektura, correction	102
korespondence, correspondence	103
korporativní autor, corporate author	100
kresba, drawing	120
kritika, critique	104
kritizovat, criticize	105
ku, to	344
kus, piece	242
kuželka, type size	354
lept, copperplate; etching	94; 137
leták, flysheet	148
letopisy *pl.*, annals *pl.*	19
lidová knihovna, public library	270
list, flysheet; leaf	148; 178
lístek, card	56
lístek katalogový, card	56
listina, document	117
lístkový katalog, card index	57
literární odkaz, literary remains	185
literatura, literature	186
litografie, lithograph(y)	187
malý, little	188
mapa, map	193
mědirytina, copperplate engraving; etching	95; 137
měnit, change[1]	63
měřítko, scale	293
městská knihovna, city library	68
mikrofilm, microfilm	200
miniatura, miniature	201
bez místa a roku (b. m. r.), no place no date	210
místo, place	243
módní časopis, fashion magazine	145
monografie, monograph(y)	204
na, on	222
náčrtek, sketch	312
nad, on	222
nadpis, title	341
náklad, edition	124
nakladatel, editor[2]; publisher[1]	126; 273
nakladatelství, publishing house	274

nákres, sketch	312
napodobení, imitation	161
národní knihovna, national library	207
nařízení, order[1]	224
nástin, sketch	312
naučný slovník, encyclopaedia	130
název, title	341
nejistý, doubtful	119
několikajazyčný, polyglot	252
nekrolog, obituary notice	216
neúplný, incomplete	165
nevydaný, unpublished	361
nezměněný, unchanged	356
normalizace, standardisation	321
notace, notation	211
novela, short story	309
noviny *pl.*, newspaper	209
nový, new	208
o, about; on	3; 222
ob, on	222
obhájit, defend	111
objednávka, order[2]	225
obraz, picture	241
obrázek, drawing	120
obsah, contents; table of contents	91; 334
obsahovat, contain	90
od, by; from; of	55; 153; 218
oddíl, paragraph; section	234; 301
odkaz, literary remains *pl.*	185
odkazový listek, reference card	278
okraj stránky, margin	194
opatřit, provide (with)	268
opis, copy[1]	96
opravit, correct	101
oprávnit, authorize	37
opravovat, correct	101
opravy, correction	102
originál, autograph	39
otisk, impression; imprint; printed matter; reprint[1]	162; 163; 262; 285
ověřený, authentic	34
padělek, forgery	151
paginace, paging	229

památník, memorial volume		199
paměti *pl.,* memoirs *pl.*		197
pamětní spis, memoir		196
papír, paper[1]		231
papyrus, papyrus		233
paragraf, paragraph		234
na **pásce,** tape, on tape		335
passim, passim		237
pergamen, parchment		235
periodický, periodical[1]		238
personální bibliografie, author bibliography		36
písemnictví, literature		186
píseň, song		314
písmo, script		299
plagiát, plagiarism		244
po, after		12
pochybný, doubtful		119
pod, under		357
na **podkladě** ..., based/founded on the ...		41
podle, after		12
podlouhlý, oblong		217
poezie, poetry		250
pohádka, fable		143
pojednání, article; treatise	31;	352
pojednávat, treat		351
pokračování, sequel		304
pokračovat, continue		92
polemický spis, polemic		251
polemika, polemic		251
polyglot, polyglot		252
polyglotní, polyglot		252
polygrafie, typography		355
popis pod názvem, title entry		342
portrét, portrait		253
posmrtný, posthumous		254
posuzovat, criticize; review *v.*	105;	288
povídka, story		324
povinná dodávka, copyright deposit		99
povinný výtisk, deposit copy		112
pozměnit, change[1]		63
poznámka, annotation; note[1]	22;	212
poznámka pod čarou, footnote		149
pramen, source		316
pravidlo, prescription; statute[1]	258;	323
president, president		259
pro, for		150
protokol, minutes *pl.*		202
próza, fiction; prose	146;	267

prozatímní lístek, temporary card		336
průběžný název, running title		292
průpověď, adage		7
průvodce, guide(-book)		155
první, first		147
první edice, original edition		227
první vydání, original edition		227
prvotisk, incunabula *pl.*		166
prvý, first		147
předběžný lístek, temporary card		336
předmětové heslo, subject heading		327
předmluva, introduction; preface	172;	256
přednáška, lecture		179
přednes, lecture		179
přední strana, recto		277
předpis, prescription; statute[1]	258;	323
předplatné, subscription		328
předseda, president		259
přehlédnout, revise[1]		289
přehlédnutí, revision		290
překlad, translation		349
překládání, translation		349
překládat, translate		348
přepracovat, rewrite		291
přetisk, reprint[1]		285
přídavek, appendix		26
příloha, supplement *n.*		331
přípisek, postscript		255
připojit, annex		20
připomínková řeč, memorial speech		198
připravit, prepare		257
připravit do tisku, edit[1]		122
příručka, handbook; pocket-book	156;	247
příručkové dílo, reference work		280
příruční knihovna, reference library		279
přispěvek, contribution		93
přizpůsobit, adapt; change[1]	8;	63
psaní, letter; script	181;	299
psát, write		372
psát básně, write poetry		373
psát verše, write poetry		373
pseudonym, pseudonym		269
původce, author		35
původní, original		226
rastr, screen		298
recenzovat, review *v.*		288
recto, recto		277

redakce, editorial office	127
redaktor, editor[1]	125
redaktorství, editorial office	127
referát, lecture; report[1]	179; 283
rejstřík, repertory; table of contents	282; 334
reklama, advertisement	11
repertorium, repertory	282
reprodukce, reproduction	286
resumé, summary	330
resumovat, summarize	329
revidovat, revise[1]	289
revize, revision	290
revue, periodical[2]; review n.	239; 287
ročenka, year-book	377
ročník, volume[2]	367
rok, volume[2]; year	367; 376
román, novel	213
romány a novely, fiction	146
rozmluva, dialog	114
rozmnožování, reproduction	286
rozprava, dialog; treatise	114; 352
rozšířit, enlarge	133
rub, verso	365
rukopis, manuscript	192
rukověť, handbook; pocket-book	156; 247
rýt, engrave	131
rytina, engraving	132
řada, series	305
řeč, speech	319
s, with	368
sazba, matter	195
sbírka, collection	75
sborník, collection	75
scénář, scenario	294
sdělení, transactions pl.	346
sebraná díla, complete works pl.	87
sebrané spisy, complete works pl.	87
sebrat, collect	74
sekce, section	301
separát, offprint	221
serie, series	305
sestavit, compile	84
sešit, booklet; number[2]	50; 215
sešívat, sew	306
seznam, list	184
shrnout, summarize	329
shrnutí, summary	330
shromáždění, assembly	32
shromáždit, collect	74
signatura, signature	310
signatura knihy, location mark	189
sjezd, assembly; congress	32; 89
skladiště knih, stack-room	320
slavostní vydání, memorial volume	199
slepecke pismo Braillovo, braille printing	52
sloka, strophe	325
sloupec, column	77
slovník, dictionary	116
služba, office	219
smíšenina, miscellanea pl.	203
smíšený, miscellanea pl.	203
souborný katalog, union catalogue	358
speciální bibliografie, special bibliography	317
speciální knihovna, special library	318
spis, document	117
spisovatel, writer	374
spojit, bind	45
společnost, society	313
spoluautor, joint author	174
spolupracovat, collaborate	72
spolupracovník, collaborator	73
spolupůsobit, collaborate	72
srovnat abecedně, alphabetize	16
starý, ancient	17
státní knihovna, state library	322
statut, statute[1]	323
století, century	62
stránka, page	228
stránkování, paging	229
strofa, strophe	325
strojopis, typescript	353
studie, study	326
svatební píseň, bridal/nuptial song	53
svazek, volume[1]	366
světlotisk, collotype	76
škola, school	295
tabule, plate	245
tabulka, table	333
těsnopis, shorthand	308
text, text	337
tisk, impression; imprint; press[1]; printed matter	162; 163; 260; 262

APPENDIX

CALENDAR UNITS — TEMPORA

leden, January	I	**prosinec,** December	XII	
únor, February	II	**pondělí,** Monday	XIII	
březen, March	III	**úterý,** Tuesday	XIV	
duben, April	IV	**středa,** Wednesday	XV	
květen, May	V	**čtvrtek,** Thursday	XVI	
červen, June	VI	**pátek,** Friday	XVII	
červenec, July	VII	**sobota,** Saturday	XVIII	
srpen, August	VIII	**neděle,** Sunday	XIX	
září, September	IX	**rok,** year	XX	
říjen, October	X	**roční,** yearly, annual	XXI	
listopad, November	XI			

šestiměsíční, half-yearly, semi-annual	XXII
tříměsíční, quarterly	XXIII
měsíc, month	XXIV
měsíční, monthly	XXV
čtrnáctidenní, fortnightly, semi-monthly	XXVI
týden, week	XXVII
týdenní, weekly	XXVIII
den, day	XXIX
denní, daily	XXX

NUMERALS — NUMERI

1	**jeden, -dna, -dno**	one	16	**šestnáct**	sixteen
2	**dva, dvě**	two	17	**sedmnáct**	seventeen
3	**tři**	three	18	**osmnáct**	eighteen
4	**čtyři**	four	19	**devatenáct**	nineteen
5	**pět**	five	20	**dvacet**	twenty
6	**šest**	six	21	**dvacet jeden**	twenty one
7	**sedm**	seven	30	**třicet**	thirty
8	**osm**	eight	40	**čtyřicet**	forty
9	**devět**	nine	50	**padesát**	fifty
10	**deset**	ten	60	**šedesát**	sixty
11	**jedenáct**	eleven	70	**sedmdesát**	seventy
12	**dvanáct**	twelve	80	**osmdesát**	eighty
13	**třináct**	thirteen	90	**devadesát**	ninety
14	**čtrnáct**	fourteen	100	**sto**	hundred
15	**patnáct**	fifteen	200	**dvě stě**	two hundred

300 tři sta	three hundred	16. šestnáctý	sixteenth
400 čtyři sta	four hundred	17. sedmnáctý	seventeenth
500 pět set	five hundred	18. osmnáctý	eighteenth
600 šest set	six hundred	19. devatenáctý	nineteenth
700 sedm set	seven hundred	20. dvacátý	twentieth
800 osm set	eight hundred	21. dvacetiprvní	twenty first
900 devět set	nine hundred	30. třicátý	thirtieth
1000 tisíc	thousand	40. čtyřicátý	fortieth
1. první	first	50. padesátý	fiftieth
2. druhý	second	60. šedesátý	sixtieth
3. třetí	third	70. sedmdesátý	seventieth
4. čtvrtý	fourth	80. osmdesátý	eightieth
5. pátý	fifth	90. devadesátý	ninetieth
6. šestý	sixth	100. stý	hundredth
7. sedmý	seventh	200. dvoustý	two hundredth
8. osmý	eighth	300. třístý	three hundredth
9. devátý	ninth	400. čtyřstý	four hundredth
10. desátý	tenth	500. pětistý	five hundredth
11. jedenáctý	eleventh	600. šestistý	six hundredth
12. dvanáctý	twelfth	700. sedmistý	seven hundredth
13. třináctý	thirteenth	800. osmistý	eight hundredth
14. čtrnáctý	fourteenth	900. devětistý	nine hundredth
15. patnáctý	fifteenth	1000. tisící	thousandth

CZECH ALPHABET

LITTERAE BOHEMICAE

A a, Á á   G g   N n, Ň ň   T t, Ť ť
B b   H h   O o, Ó ó   U u, Ú ú, ů
C c   Ch ch   P p   V v
Č č   I i, Í í   Q q   W w
D d, Ď ď   J j   R r   X x
E e, É é, Ě ě   K k   Ř ř   Y y ý
F f   L l   S s   Z z
  M m   Š š   Ž ž

DANISH — DANICA

blindeskrift, braille printing	52
bo, literary remains *pl.*	185
bog, book	49
bogbinderi, bindery	46
boghandel, bookshop	51
bogkundskab, bibliology	43
bogmagasin, stack-room	320
bogtrykker, printer	264
bogtrykkeri, printing office	266
brev, letter	181
brevveksling, correspondence	103
brochure, booklet	50
brudstykke, fragment	152
brug, use *n.*	362
bruge, use *v.*	363
bryllupsdigt, bridal/nuptial song	53
bulletin, bulletin	54
børnebog, children's book	67
på bånd, tape, on tape	335
chef, chief	66
dagblad, newspaper	209
dagbog, diary	115
de *pl.*, the	339
decimalklassifikation, decimal classification	107
dediceret, dedicated	108
dedikation, dedication	109
del, part; section	236; 301
den, the	339
desiderata, desiderata	113
det, the	339
dialog, dialog	114
digt, poem	248
digte, write poetry	373
digter, poet	249
digtning, poem	248
disputats, thesis	340
dokument, document	117
dokumentation, documentation	118
drage omsorg for, edit[1]	122
drama, play	246
drejebog, scenario	294
duplikat, duplicate	121
efter, after	12
efterladenskab, literary remains *pl.*	185
efterladt, posthumous	254

efterligning, imitation	161
efterskrift, epilog; postscript	135; 255
eksemplar, copy[2]	97
elegie, elegy	129
embede, office	219
en, a, an	1
encyklopædi, encyclopaedia	130
eneste, only	223
enestående, only	223
epilog, epilogue	135
epistel, letter	181
epope, epic	134
erindringer *pl.*, memoirs *pl.*	197
essay, essay[1]; study	136; 326
et, a, an	1
eventyr, fable	143
fabel, fable	143
fagbibliografi, special bibliography	317
fagbibliotek, special library	318
faksimile, facsimile	144
falskneri, forgery	151
figur, picture	241
flersproget, polyglot	252
flyveblad, flysheet	148
flyveskrift, flysheet	148
folkebibliotek, public library	270
for, for	150
forandre, change[1]	63
forandre sig, change[2]	64
forbedre, correct	101
forberede, prepare	257
foredrag, lecture	179
forelæsning, lecture	179
foreløbig seddel, temporary card	336
foreløbigt kort, temporary card	336
forfalskning, forgery	151
forfatter, author; writer	35; 374
forfatternavn, pseudonym	269
forhandlinger, minutes *pl.*	202
forklare, explain	140
forkorte, abridge	4
forkortet indførsel, abbreviated entry	2
forlag, publishing house	274
forlægge, publish	271
forlægger, publisher[1]	273
forordning, order[1]	224
forsamling, assembly	32

konversationsleksikon, encyclopaedia	130	
kopi, copy[1]	96	
korporativ forfatter, corporate author	100	
korrektur, correction	102	
korrespondance, correspondence	103	
kort, card	56	
krestomati, selection	303	
kritik, critique	104	
kritisere, criticize	105	
kunst, art	30	
landkort, map	193	
leksikon, encyclopaedia	130	
levende kolumnetitel, running title	292	
levnedsbeskrivelse, biography	48	
lille, little	188	
liste, list	184	
litografi, lithograph(y)	187	
litteratur, literature	186	
lov, law	177	
lovsang, panegyric	230	
lystryk, collotype	76	
lystspil, comedy	78	
lærebog, textbook	338	
løsbladsbog, loose-leaf book	190	
manuskript, manuscript	192	
margin, margin	194	
maskinskrevet eksemplar, typescript	353	
maskinskrevet manuskript, typescript	353	
med, with	368	
medarbejder, collaborator	73	
meddelelse, transactions pl.	346	
medforfatter, joint author	174	
medvirke, collaborate	72	
meldinger pl., transactions pl.	346	
memoirer pl., memoirs pl.	197	
mikrofilm, microfilm	200	
mindeskrift, memoir; memorial volume	196; 199	
mindetale, memorial speech	198	
miniatur, miniature	201	
modeblad, fashion magazine	145	
modejournal, fashion magazine	145	
monografi, monograph(y)	204	
musik, music	205	
musikalier pl., printed music	263	
møde, assembly; sitting	32; 311	
målestok, scale	293	

nationalbibliotek, national library	207	
navn, name	206	
nekrolog, obituary notice	216	
noder pl., printed music	263	
notation, notation	211	
note, annotation; footnote	22; 149	
novelle, short story; story	309; 324	
nummer, number[1]; number[2]	214; 215	
ny, new	208	
officiel, official	220	
og, and	18	
om, about; on	3; 222	
omarbejde, rewrite	291	
omdanne, change[1]	63	
ophav, source	316	
ophavsret, copyright	98	
ophørt at udkomme, ceased publication	61	
oplag, edition; impression	124; 162	
optryk, reprint[1]	285	
ordbog, dictionary	116	
ordliste, glossary	154	
ordne, arrange	29	
ordre, order[1]	224	
ordstrid, polemic	251	
original, original	226	
originalhåndskrift, autograph	39	
originaludgave, original edition	227	
over, about; on	3; 222	
oversætte, translate	348	
oversættelse, translation	349	
paginering, paging	229	
papir, paper[1]	231	
papyrus, papyrus	233	
paragraf, paragraph	234	
passim, passim	237	
pergament, parchment	235	
periodisk, periodical[1]	238	
personalbibliografi, author bibliography	36	
pladssignatur, location mark	189	
plagiat, plagiarism	244	
planche, plate	245	
pligtaflevering, copyright deposit	89	
pligtafleveringseksemplar, deposit copy	112	
poesi, poetry	250	

tekst, text	337
tidsskrift, periodical[2]	239
til, for; to	150; 344
tilegnelse, dedication	109
tillempe, adapt	8
tillæg, addenda *pl.*; appendix; supplement *n.*	10; 26; 331
tillægs-, supplementary	332
titel, title	341
titelblad, title page	343
tosproget, bilingual	44
tragedie, tragedy	345
traktat, treatise	352
translitteration, transliteration	350
transskription, transcription	347
tryk, impression	162
trykfejl, printer's error	265
trykke, print[1]	261
trykkeri, printing office	266
tryksag, impression; printed matter	162; 262
trykår, date of printing	106
træsnit, woodcut	370
træstik, woodcut	370
tvivlsom, doubtful	119
i tværformat, oblong	217
typografi, typography	355
udarbejde, elaborate	128
uddrag, extract	142
uden, without	369
udgave, edition	124
udgive, edit[2]	123
udgiver, editor[2]	126
udkast, sketch	312

udkomme, *be* published	272
udlån, lending	180
udstilling, exhibition	139
udvekslingstjeneste, exchange centre	138
udvide, enlarge	133
udvælge, select	302
uforandret, unchanged	356
ufuldstændig, defective; incomplete	110; 165
ukomplet, defective; incomplete	110; 165
under, under	357
universitet, university	359
universitetsbibliotek, university library	360
u. s. o. å., → *uden* sted og år	
utrykt, unpublished	361
variant, version	364
vejledning, guide(-book)	155
version, version	364
videnskab, science	296
vittighedsblad, comic paper	79
vuggetryk, incunabula *pl.*	166
vælge, select	302
værk, work	371
xerografi, xerography	375
ægte, authentic	34
år, year	376
årbog, year-book	377
årgang, volume[2]	367
århundrede, century	62

APPENDIX

CALENDAR UNITS — TEMPORA

januar, January	I	**december,** December	XII	**halvårlig,** half-yearly,	
februar, February	II	**mandag,** Monday	XIII	semi-annual	XXII
marts, March	III	**tirsdag,** Tuesday	XIV	**kvartals-,** quarterly	XXIII
april, April	IV	**onsdag,** Wednesday	XV	**måned,** month	XXIV
maj, May	V	**torsdag,** Thursday	XVI	**månedlig,** monthly	XXV
juni, June	VI	**fredag,** Friday	XVII	**halvmånedlig,** fortnightly,	
juli, July	VII	**lørdag,** Saturday	XVIII	semi-monthly	XXVI
august, August	VIII	**søndag,** Sunday	XIX	**uge,** week	XXVII
september, September	IX	**år,** year	XX	**ugentlig,** weekly	XXVIII
oktober, October	X	**årlig,** annual, yearly	XXI	**dag,** day	XXIX
november, November	XI			**daglig,** daily	XXX

NUMERALS — NUMERI

1 **een, eet**	one	16 **seksten**	sixteen
2 **to**	two	17 **sytten**	seventeen
3 **tre**	three	18 **atten**	eighteen
4 **fire**	four	19 **nitten**	nineteen
5 **fem**	five	20 **tyve**	twenty
6 **seks**	six	21 **en ogtyve**	twenty-one
7 **syv**	seven	30 **tredive**	thirty
8 **otte**	eight	40 **fyrre, fyrretyve**	forty
9 **ni**	nine	50 **halvtreds(indstyve)**	fifty
10 **ti**	ten	60 **tres(indstyve)**	sixty
11 **elleve**	eleven	70 **halvfjerds(indstyve)**	seventy
12 **tolv**	twelve	80 **firs(indstyve)**	eighty
13 **tretten**	thirteen	90 **halvfems(indstyve)**	ninety
14 **fjorten**	fourteen	100 **hundrede**	hundred
15 **femten**	fifteen	200 **to hundrede**	two hundred

265

300 tre hundrede	three hundred	16. sekstende	sixteenth
400 fire hundrede	four hundred	17. syttende	seventeenth
500 fem hundrede	five hundred	18. attende	eighteenth
600 seks hundrede	six hundred	19. nittende	nineteenth
700 syv hundrede	seven hundred	20. tyvende	twentieth
800 otte hundrede	eight hundred	21. en og tyvende	twenty-first
900 ni hundrede	nine hundred	30. tredivte	thirtieth
1000 tusind(e)	thousand	40. fyrretyvende	fortieth
1. første	first	50. halvtredsindstyvende	fiftieth
2. anden	second	60. tresindstyvende	sixtieth
3. tredje	third	70. halvfjerdsindstyvende	seventieth
4. fjerde	fourth	80. firsindstyvende	eightieth
5. femte	fifth	90. halvfemsindstyvende	ninetieth
6. sjette	sixth	100. hundrede	hundredth
7. syvende	seventh	200. to hundrede	two hundredth
8. ottende	eighth	300. tre hundrede	three hundredth
9. niende	ninth	400. fire hundrede	four hundredth
10. tiende	tenth	500. fem hundrede	five hundredth
11. ellevte	eleventh	600. seks hundrede	six hundredth
12. tolvte	twelfth	700. syv hundrede	seven hundredth
13. trettende	thirteenth	800. otte hundrede	eight hundredth
14. fjortende	fourteenth	900. ni hundrede	nine hundredth
15. femtende	fifteenth	1000. tusinde	thousandth

DANISH ALPHABET

LITTERAE DANICAE

A a	K k	U u
B b	L l	V v
C c	M m	W w
D d	N n	X x
E e	O o	Y y
F f	P p	Z z
G g	Q q	Æ æ
H h	R r	Ø ø
I i	S s	Å å
J j	T t	

DUTCH — HOLLANDICA

plaat, plate	245
plaats, place	243
zonder **plaats en jaar (z. pl. e. j.)** no place no date	210
plagiaat, plagiarism	244
poëzie, poetry	250
polemiek, polemic	251
portret, portrait	253
postuum, posthumous	254
prent, engraving	132
presentexemplaar, deposit copy	112
president, president	259
proefschrift, thesis	340
protocol, minutes *pl.*	202
proza, prose	267
pseudoniem, pseudonym	269
publiceren, edit[2]	123
rangschikken, arrange	29
rapport, report[1]	283
raster, screen	298
recenseren, review *v.*	288
reclame, advertisement	11
recto, recto	277
redacteur, editor[1]	125
redactie, editorial office	127
rede, speech	319
reeks, series	305
regel, prescription; statute[1]	258; 323
reglement, prescription	258
repertorium, repertory	282
reproductie, reproduction	286
revue, review *n.*	287
roman, novel	213
romanliteratuur, fiction	146
rubricering, classification	69
rug, back	40
ruilbureau, exchange centre	138
ruildienst, exchange centre	138
samenspraak, dialog	114
samenstellen, compile	84
samenvatten, summarize	329
samenvatting, summary	330
samenwerken, collaborate	72
schaal, scale	293
scheets, sketch	312

school, school	295
schoolboek, textbook	338
schrift, script	299
schrijven, write	372
schrijver, author; writer	35; 374
sectie, section	301
serie, series	305
signatuur, signature	310
slotwoord, epilogue	135
speciale bibliografie, special bibliography	317
speciale bibliotheek, special library	318
spraak, language	175
spreekwoord, adage	7
staatsbibliotheek, state library	322
stadsbibliotheek, city library	68
standardisatie, standardisation	321
statuut, statute[1]	323
steendruk, lithograph(y)	187
stenografie, shorthand	308
strofe, strophe	325
studie, study	326
stuk, piece	242
supplement, addenda *pl.*	10
supplement-, supplementary	332
taal, language	175
tabel, list; table	184; 333
tafel, plate	245
te, to	344
tegenspraak, refutation	281
tekening, drawing	120
tekst, text	337
tentoonstelling, exhibition	139
tijdschrift, periodical[2]	239
titel, title	341
titelbeschrijving, cataloging	60
titelblad, title page	343
titelkaart, title entry	342
titelpagina, title page	343
toegepast, applied	27
tokomstroman, science fiction	297
toneelstuk, play	246
tot, to	344
tragedie, tragedy	345
tractaat, treatise	352
transcriptie, transcription	347
transliteratie, transliteration	350

APPENDIX

CALENDAR UNITS — TEMPORA

januari, January	I	**maandag,** Monday	XIII	**driemaandelijks,** quarterly	XXIII
februari, February	II	**dinsdag,** Tuesday	XIV	**maand,** month	XXIV
maart, March	III	**woensdag,** Wednesday	XV	**maandelijks,** monthly	XXV
april, April	IV	**donderdag,** Thursday	XVI	**halfmaandelijks,**	
mei, May	V	**vrijdag,** Friday	XVII	fortnightly, semi-	
juni, June	VI	**zaterdag,** Saturday	XVIII	monthly	XXVI
juli, July	VII	**zondag,** Sunday	XIX	**week,** week	XXVII
augustus, August	VIII	**jaar,** year	XX	**weekelijks,** weekly	XXVIII
september, September	IX	**jaarlijks,** yearly	XXI	**dag,** day	XXIX
october, October	X	**halfjaarlijks,** half-yearly,		**dagelijks,** daily	XXX
november, November	XI	semi-annual	XXII		
december, December	XII				

NUMERALS — NUMERI

1 **een**	one		16 **zestien**	sixteen
2 **twee**	two		17 **zeventien**	seventeen
3 **drie**	three		18 **achttien**	eighteen
4 **vier**	four		19 **negentien**	nineteen
5 **vijf**	five		20 **twintig**	twenty
6 **zes**	six		21 **eenentwintig**	twenty one
7 **zeven**	seven		30 **dertig**	thirty
8 **acht**	eight		40 **veertig**	forty
9 **negen**	nine		50 **vijftig**	fifty
10 **tien**	ten		60 **zestig**	sixty
11 **elf**	eleven		70 **zeventig**	seventy
12 **twaalf**	twelve		80 **tachtig**	eighty
13 **dertien**	thirteen		90 **negentig**	ninety
14 **veertien**	fourteen		100 **honderd**	hundred
15 **vijftien**	fifteen		200 **tweehonderd**	two hundred

300	driehonderd	three hundred	
400	vierhonderd	four hundred	
500	vijfhonderd	five hundred	
600	zeshonderd	six hundred	
700	zevenhonderd	seven hundred	
800	achthonderd	eight hundred	
900	negenhonderd	nine hundred	
1000	duizend	thousand	
1.	eerste	first	
2.	tweede	second	
3.	derde	third	
4.	vierde	fourth	
5.	vijfde	fifth	
6.	zesde	sixth	
7.	zevende	seventh	
8.	achtste	eighth	
9.	negende	ninth	
10.	tiende	tenth	
11.	elfde	eleventh	
12.	twaalfde	twelfth	
13.	dertiende	thirteenth	
14.	veertiende	fourteenth	
15.	vijftiende	fifteenth	

16.	zestiende	sixteenth	
17.	zeventiende	seventeenth	
18.	achttiende	eighteenth	
19.	negentiende	nineteenth	
20.	twintigste	twentieth	
21.	eenentwintigste	twenty first	
30.	dertigste	thirtieth	
40.	veertigste	fortieth	
50.	vijftigste	fiftieth	
60.	zestigste	sixtieth	
70.	zeventigste	seventieth	
80.	tachtigste	eightieth	
90.	negentigste	ninetieth	
100.	honderdste	hundredth	
200.	tweehonderdste	two hundredth	
300.	driehonderdste	three hundredth	
400.	vierhonderdste	four hundredth	
500.	vijfhonderdste	five hundredth	
600.	zeshonderdste	six hundredth	
700.	zevenhonderdste	seven hundredth	
800.	achthonderdste	eight hundredth	
900.	negenhonderdste	nine hundredth	
1000.	duizendste	thousandth	

FINNISH – FENNICA

esitellä, annotate	21
esitelmä, lecture	179
esittää, represent	284
essee, essay[1]	136
etsaus, etching	137
faksimile, facsimile	144
falsifikaatti, forgery	151
fragmentti, fragment	152
hajakohdin, passim	237
hakemisto, index	167
hakusana, subject heading	327
hakuteos, reference work	280
harvinainen, rare	275
henkilöbibliografia, author bibliography	36
historia, history	157
huomautus, note[1]	212
huvinäytelmä, comedy	78
hyllysignumi, location mark	189
hymni, anthem	24
häälaulu, bridal/nuptial song	53
idylli, idyll	159
ikonografia, iconography	158
illustraatio, picture	241
ilman, without	369
ilmestymästä lakannut julkaisu, ceased publication	61
ilmestyä, be published	272
ilmoitus, advertisement; report[1]; transactions pl.	11; 283; 346
initiaalit, pl. initial	168
inkunaabeli, incunabula pl.	166
instituutti, institute	170
inventaari(o), inventory	173
i. p. & v., → ilman painopaikkaa ja vuotta	
irtolehtikirja, loose-leaf book	190
iso, large	176
istunto, sitting	311
ja, and	18
jakso, series	305
jaosto, section	301
jatkaa, continue	92
jatko, sequel	304
johdanto, introduction; preface	172; 256
johdatta, introduce	171
juhlakirja, memorial volume	199
julkaisematon, unpublished	361
julkaisija, editor[2]	126
julkaista, edit[2]	123
julkaisu, edition	124
julkaisutiedot, imprint	163
jäljennys, reproduction	286
jäljennös, copy[1]	96
jäljennöspainos, facsimile	144
jäljittely, imitation	161
jälkeen, after	12
jälkikirjoitus, postscript	255
jälkilause, epilogue	135
jälkipainos, reprint[1]	285
jälkipuhe, postscript	255
järjestää, arrange; edit[1]	29; 122
jäännös, literary remains pl.	185
kaiverrus, engraving	132
kaivertaa, engrave	131
kaksikielinen, bilingual	44
kaksoiskappale, duplicate	121
kansalliskirjasto, national library	207
kanssa, with	368
kantapainos, original edition	227
kappale, chapter; copy[2]; piece	65; 97; 242
kartasto, atlas	33
kartta, map	193
katkelma, fragment	152
katsaus, review n.	287
kaunokirjallisuus, literature	186
kaupunginkirjasto, city library	68
kera, with	368
kertoelma, story	324
kertomakirjallisuus, fiction	146
kertomus, history; report[1]; short story; story	157; 283; 309; 324
kerätä, collect	74
keskustella, treat	351
keskustelu, treatise	352
kieli, language	175
kiinnittää, sew	306
kirja, book	49
kirjailija, writer	374

-lla, with	368
-lle, for; to	150; 344
-llä, with	368
loppulause, epilogue	135
-lta, from	153
-ltä, from	153
luettelo, catalog(ue); list	58; 184
luetteloida, catalog(ue)	59
luettelointi, cataloging	60
luettelokortti, card	56
luku, number[1]	214
luokanmerkki, signature	310
luokitus, classification	69
luo(kse), to	344
luonnos, sketch	312
lyhennetty kirjaus, abbreviated entry	2
lyhentää, abridge	4
lähde, source	316
mainos, advertisement	11
marginaali, margin	194
memoaarit *pl.,* memoirs *pl.*	197
merkistö, notation	211
merkkijärjestelmä, notation	211
mietelause, adage	7
mietelmä, adage	7
mikrofilmi, microfilm	200
mittakaava, scale	293
monikielinen, polyglot	252
monistus, reproduction	286
monografia, monograph(y)	204
muinainen, ancient	17
muistelmat *pl.,* memoirs *pl.*	197
muistojulkaisu, memorial volume	199
muistokirjoitus, memoir; obituary notice	196; 216
muistutus, note[1]	212
mukailu, imitation	161
mukana, with	368
muokata, elaborate; rewrite	128; 291
muotilehti, fashion magazine	145
muotokuva, portrait	253
murhenäytelmä, tragedy	345
musiikki, music	205
muunnos, version	364
muuntaa, change[1]	63
muuntua, change[2]	64
muuttamaton, unchanged	355

muuttua, change[2]	64
määräys, order[1]	224
-n, by; of	55; 218
nauhaan, on tape	335
nauhalle, on tape	335
nekrologi, obituary notice	216
nide, volume[1]	366
nidos, volume[1]	366
nidottu kirja, paperback	232
nimekkeenmukainen kirjaus, title entry	342
nimetön, anonymous	23
nimi, name; title	206; 341
nimiölehti, title page	343
nitoa, sew	306
normaalistaminen, standardisation	321
novelli, short story	309
numero, number[1]; number[2]	214; 215
nuotit *pl.,* printed music	263
näytelmä, play	246
näyttely, exhibition	139
näytös, act	6
ohje, prescription; statute[1]	258; 323
oikaista correct	101
oikaisu, correction	102
oikeaperäinen, authentic	34
oikeuttaa, authorize	37
omaelämäkerta, autobiography	38
omistettu, dedicated	108
omistus, dedication	109
omistuskirjoitus, dedication	109
opas, guide(-book)	155
opisto, institute	170
oppikirja, textbook	338
osa, part	236
osasto, section	301
ote, extract	142
otsikko, title	341
paikanmerkki, location mark; signature	189; 310
paikka, place	243
paikkamerkki, location mark	189
painaa, print[1]	261
painaja, printer	264

-ssa, in 164
-ssä, in 164
-sta, about; from; on 3; 153; 222
stroofi, strophe 325
-stä, about; from; on 3; 153; 222
suorasanainen kaunokirjallisuus, fiction 146
surulaulu, elegy 129
suunnitelma, sketch 312
suuri, large 176
syövytys, etching 137
säkeistö, strophe 325
sävelmäjulkaisu, printed music 263
säädös, order[1] 224
sääntö, prescription; statute[1] 258; 323

taide, art 30
takia, for 150
tarina, fable 143
tarkastaa, revise[1] 289
tarkastus, revision 290
tarkistus, revision 290
taskukirja, pocket-book 247
taulu, plate 245
taulukko, table 333
tekijä, author; writer 35; 374
tekijäkumppani, joint author 174
tekijänoikeus, copyright 98
teksti, text 337
teos, work 371
tiede, science 296
tiedonanto, transactions pl. 346
tiedonanto(lehti), bulletin 54
tiedotus, transactions pl. 346
tiedotuslehti, bulletin 54
tieteisromaani, science fiction 297
tietosanakirja, encyclopaedia 130
tilapäiskortti, temporary card 336
tilapäislippu, temporary card 336
tilaus, order[2] 225
titteli, title 341
toimia yhdessä, collaborate 72
toimikunta, committee 83
toimittaja, editor[1] 125
toimitus, editorial office 127
tragedia, tragedy 345
translitteraatio, transliteration 350
translitterointi, transliteration 350

tutkielma, essay[1]; study; treatise 136; 326; 352
typografia, typography 355
täydennys, addenda pl. 10
täydentää, complete 86

useissa kohdin, passim 237
uusi, new 208
uusia, rewrite 291

vaihtokeskus, exchange centre 138
vaihtuva sivu(n)otsikko, running title 292
valaista, illustrate 160
valikoida, select 302
valita, select 302
valitusruno, elegy 129
valmistaa, prepare 257
valmistella, prepare 257
valokuva, photograph 240
valopainate, collotype 76
valtionkirjasto, state library 322
valtuuttaa, authorize 37
vanha, ancient 17
vapaakappale, deposit copy 112
varhaispainos, incunabula pl. 166
varten, for 150
varustaa, provide (with) 268
vaskipiirros, copperplate engraving 95
vastatodistus, refutation 281
vedos, impression 162
vierus, margin 194
vihko, booklet 50
viitekortti, reference card 278
viittauskortti, reference card 278
virallinen, official 220
virasto, office 219
virka, office 219
vuoksi, for 150
vuoropuhelu, dialog 114
vuosi, year 376
vuosikerta, volume[2] 367
vuosikirja, year-book 377
vuosikirjat pl., annals pl. 19
vuosisata, century 62
väittelykirjoitus, polemic 251
väitöskirja, thesis; treatise 340; 352
väliaikaiskortti, temporary card 336

APPENDIX

CALENDAR UNITS — TEMPORA

tammikuu, January	I	maanantai, Monday	XIII	kolmen kuukauden, quarterly	XXIII
helmikuu, February	II	tiistai, Tuesday	XIV	kuukausi, month	XXIV
maaliskuu, March	III	keskiviikko, Wednesday	XV	jokakuukautinen, monthly	XXV
huhtikuu, April	IV	torstai, Thursday	XVI	puolikuukautinen, fort-	
toukokuu, May	V	perjantai, Friday	XVII	nightly, semi-monthly	XXVI
kesäkuu, June	VI	lauantai, Saturday	XVIII	viikko, week	XXVII
heinäkuu, July	VII	sunnuntai, Sunday	XIX	viikko-, weekly	XXVIII
elokuu, August	VIII	vuosi, year	XX	päivä, day	XXIX
syyskuu, September	IX	vuotinen, annual, yearly	XXI	päivä-, daily	XXX
lokakuu, October	X	puolivuotinen, half-yearly,			
marraskuu, November	XI	semi-annual	XXII		
joulukuu, December	XII				

NUMERALS — NUMERI

1	yksi	one	16	kuusitoista	sixteen
2	kaksi	two	17	seitsemäntoista	seventeen
3	kolme	three	18	kahdeksantoista	eighteen
4	neljä	four	19	yhdeksäntoista	nineteen
5	viisi	five	20	kaksikymmentä	twenty
6	kuusi	six	21	kaksikymmentäyksi	twenty one
7	seitsemän	seven	30	kolmekymmentä	thirty
8	kahdeksan	eight	40	neljäkymmentä	forty
9	yhdeksän	nine	50	viisikymmentä	fifty
10	kymmenen	ten	60	kuusikymmentä	sixty
11	yksitoista	eleven	70	seitsemänkymmentä	seventy
12	kaksitoista	twelve	80	kahdeksankymmentä	eighty
13	kolmetoista	thirteen	90	yhdeksänkymmentä	ninety
14	neljätoista	fourteen	100	sata	hundred
15	viisitoista	fifteen	200	kaksisataa	two hundred

300	kolmesataa	three hundred	
400	neljäsataa	four hundred	
500	viisisataa	five hundred	
600	kuusisataa	six hundred	
700	seitsemänsataa	seven hundred	
800	kahdeksansataa	eight hundred	
900	yhdeksänsataa	nine hundred	
1000	tuhat	thousand	
1.	ensimmäinen	first	
2.	toinen	second	
3.	kolmas	third	
4.	neljäs	fourth	
5.	viides	fifth	
6.	kuudes	sixth	
7.	seitsemäs	seventh	
8.	kahdeksas	eighth	
9.	yhdeksäs	ninth	
10.	kymmenes	tenth	
11.	yhdestoista	eleventh	
12.	kahdestoista	twelfth	
13.	kolmastoista	thirteenth	
14.	neljästoista	fourteenth	
15.	viidestoista	fifteenth	

16.	kuudestoista	sixteenth
17.	seitsemästoista	seventeenth
18.	kahdeksastoista	eighteenth
19.	yhdeksästoista	nineteenth
20.	kahdeskymmenes	twentieth
21.	kahdeskymmenesyhdes	twenty first
30.	kolmaskymmenes	thirtieth
40.	neljäskymmenes	fortieth
50.	viideskymmenes	fiftieth
60.	kuudeskymmenes	sixtieth
70.	seitsemäskymmenes	seventieth
80.	kahdeksaskymmenes	eightieth
90.	yhdeksäskymmenes	ninetieth
100.	sadas	hundredth
200.	kahdessadas	two hundredth
300.	kolmassadas	three hundredth
400.	neljässadas	four hundredth
500.	viidessadas	five hundredth
600.	kuudessadas	six hundredth
700.	seitsemässadas	seven hundredth
800.	kahdeksassadas	eight hundredth
900.	yhdeksässadas	nine hundredth
1000.	tuhannes	thousandth

FINNISH ALPHABET

LITTERAE FENNICAE

A a	J j	S s
B b	K k	T t
C c	L l	U u
D d	M m	V v
E e	N n	W w
F f	O o	Y y
G g	P p	Z z
H h	Q q	Ä ä
I i	R r	Ö ö

GREEK – GRAECA

ek *(ἐκ)*, from; of   153; 218
ekdido *(ἐκδίδω)*, edit²; publish   123; 271
ekdidomai *(ἐκδίδομαι)*, be published   272
ekdosis *(ἔκδοσις)*, edition   124
ekdosis prote *(ἔκδοσις πρώτη)*, original edition   227
ekdotes *(ἐκδότης)*, editor²; publisher¹   126; 273
ekdotikos oikos *(ἐκδοτικός οἶκος)*, publishing house   274
eklego *(ἐκλέγω)*, select   302
ektaktos anatyposis *(ἔκτακτος ἀνατύπωσις)*, offprint   221
ekthesis *(ἔκθεσις)*, exhibition   139
ektypono *(ἐκτυπώνω)*, print¹   261
ektyponomai *(ἐκτυπώνομαι)*, be published   272
ektyposis *(ἐκτύπωσις)*, imprint   163
elattomatikos *(ἐλαττωματικός)*, defective   110
elegeia *(ἐλεγεία)*, elegy   129
ell(e)ipes *(ἐλλειπής)*, incomplete   165
elencho *(ἐλέγχω)*, revise¹   289
elenchos *(ἔλεγχος)*, list   184
en *(ἐν)*, in   164
enarmozo *(ἐναρμόζω)*, adapt   8
encheiridion *(ἐγχειρίδιον)*, handbook; pocket-book;
    textbook   156; 247; 338
engrafa *(ἔγγραφα) pl.*, document   117
ep'eniauton ephemerides *(ἐπ᾿ἐνιαυτόν ἐφημερίδες)*,
    volume²   367
enkarsios *(ἐγκάρσιος)*, oblong   217
enkyklopaideia *(ἐγκυκλοπαίδεια)*, encyclopaedia   130
entypon *(ἔντυπον)*, printed matter   262
epeteios *(ἐπέτειος)*, volume²   367
epeteris *(ἐπετηρίς)*, year-book   377
ephemeris *(ἐφημερίς)*, newspaper   209
ephermosmenos *(ἐφηρμοσμένος)*, applied   27
ephodiazo *(ἐφοδιάζω)*, provide (with)   268
epi *(ἐπί)*, to   344
epicheirematologia *(ἐπιχειρηματολογία)*,
    documentation   118
epikedeios logos *(ἐπικήδειος λόγος)*, memorial
    speech   198
epikrino *(ἐπικρίνω)*, criticize   105
epilogos *(ἐπίλογος)*, epilogue   135
epimekes *(ἐπιμήκης)*, oblong   217
epimeleomai *(ἐπιμελέομαι)*, edit¹   122
episema *(ἐπίσημα)*, location mark   189
episemos *(ἐπίσημος)*, official   220
episteme *(ἐπιστήμη)*, science   296
epistemonikon phantastikon mythistorema
    (ἐπιστημονικόν φανταστικόν μυθιστόρημα),
    science fiction   297

epistole *(ἐπιστολή)*, letter   181
epithalamion *(ἐπιθαλάμιον)*, bridal song   53
epitheoro *(ἐπιθεωρῶ)*, revise¹   289
epititlos selidos *(ἐπίτιτλος σελίδος)*, running title   292
epitome *(ἐπιτομή)*, extract   142
epitrepo *(ἐπιτρέπω)*, authorize   37
epitrope *(ἐπιτροπή)*, committee   83
epitymbidion *(ἐπιτυμβίδιον)*, elegy   129
epopoiia *(ἐποποιΐα)*, epic   134
epos *(ἔπος)*, epic   134
ergon *(ἔργον)*, work   371
esokleio *(ἐσωκλείω)*, annex   20
ethnike bibliotheke *(ἐθνική βιβλιοθήκη)*, national
    library   207
etos *(ἔτος)*, year   376
euryno *(εὐρύνω)*, enlarge   133
ex *(ἐξ)*, of   218
exegesis *(ἐξήγησις)*, commentary; explanation   82; 141
exego *(ἐξηγῶ)*, comment¹; explain   81; 140
exergazomai *(ἐξεργάζομαι)*, elaborate   128
exusian parecho *(ἐξουσίαν παρέχω)*, authorize   37

faksimile *(φαξίμιλε)*, facsimile   144
feig-bolan *(φέϊγ-βολάν)*, flysheet   148

gamelia *(γαμελία)*, bridal song   53
geographikos chartes *(γεωγραφικός χάρτης)*, map   193
gia *(γιά)*, about; for; on   3; 150; 222
glossa *(γλῶσσα)*, language   175
glotta *(γλῶττα)*, language   175
glyphe *(γλύφη)*, engraving   132
glypho *(γλύφω)*, engrave   131
gnome *(γνώμη)*, adage   7
grammata *(γράμματα) pl.*, correspondence   103
graphe *(γραφή)*, script   299
graphe Braille *(γραφή Braille)*, braille printing   52
grapho *(γράφω)*, write   372
grapho poiemata *(γράφω ποιήματα)*, write poetry   373

hai *(αἱ) pl.*, the   339
hapanta *(ἄπαντα)*, complete works   87
he *(ἡ)*, the   339
heis *(εἷς)*, a, an   1
hekastachu *(ἑκασταχοῦ)*, passim   237
hekatontaeteris *(ἑκατονταετηρίς)*, century   62

metathanatios *(μεταθανάτιος)*, posthumous 254
metatyposis *(μετατύπωσις)*, reprint[1] 285
mia *(μία)*, a, an 1
mikrographia *(μικρογραφία)*, miniature 201
mikrophilm *(μικροφίλμ)*, microfilm 200
mikros *(μικρός)*, little 188
mimesis *(μίμησις)*, imitation 161
miniatura *(μινιατούρα)*, miniature 201
monografia *(μονογραφία)*, monograph(y) 204
monos *(μόνος)*, only 223
mprosura *(μπροσούρα)*, booklet 50
musike *(μουσική)*, music 205
mythistorema *(μυθιστόρημα)*, novel 213
mythistorema epistemonikes phantasias
    (μυθιστόρημα ἐπιστημονικῆς φαντασίας),
    science fiction 297
mythos *(μῦθος)*, fable 143

nekrologion *(νεκρολόγιον)*, obituary notice 216
neos *(νέος)*, new 208
kata nomon prosphora *(κατά νόμον προσφορά)*
    copyright deposit 99
nomos *(νόμος)*, law 177
notheusis *(νόθευσις)*, forgery 151
numero *(νούμερο)*, number[2] 215

ode *(ὠδή)*, song 314
onoma *(ὄνομα)*, name 206
onomastikon *(ὀνομαστικόν)*, glossary 154

paidikon biblion *(παιδικόν βιβλίον)*, children's
    book 67
palaiobibliopoleion *(παλαιοβιβλιοπωλεῖον)*,
    second-hand bookshop 300
pal(a)ios *(παλαιός)*, ancient 17
panegyrikos (logos) *(πανηγυρικός λόγος)*, panegyric 230
panepistemiake bibliotheke *(πανεπιστημιακή
    βιβλιοθήκη)*, university library 360
panepistemion *(πανεπιστήμιον)*, university 359
pantachu *(πανταχοῦ)*, passim 237
papyros *(πάπυρος)*, papyrus 233
paradosis *(παράδοσις)*, lecture 179
paragraphos *(παράγραφος)*, paragraph 234
paramythi *(παραμύθι)*, fable 143
parangelia *(παραγγελία)*, order[2] 225
parapompe *(παραπομπή)*, footnote 149

parartema *(παράρτημα)*, appendix;
    supplement *n.* 26; 331
paraskeuazo *(παρασκευάζω)*, prepare 257
parecho *(παρέχω)*, provide (with) 268
paremballo *(παρεμβάλλω)*, insert 169
parergon *(πάρεργον)*, appendix 26
partitura *(παρτιτούρα)*, printed music 263
pege *(πηγή)*, source 316
peri *(περί)*, about 3
periecho *(περιέχω)*, contain 90
periechomenon *(περιεχόμενον)*, contents 91
perigrapho *(περιγράφω)*, represent 284
perilepsis *(περίληψις)*, summary 330
periodikon *(περιοδικόν)*, periodical[2] *n.* 239
periodikon modas *(περιοδικόν μόδας)*, fashion
    magazine 145
periodikos *(περιοδικός)*, periodical[1] *a.* 238
perithorion *(περιθώριον)*, margin 194
pergamene *(περγαμηνή)*, parchment 253
pezos logos *(πεζός λόγος)*, prose 267
philologia *(φιλολογία)*, literature 186
sten phonotainian *(στήν φωνοταινίαν)*, on tape 335
photographia *(φωτογραφία)*, photograph 240
phototypia *(φωτοτυπία)*, collotype 76
phylladion *(φυλλάδιον)*, booklet 50
phyllon *(φύλλον)*, leaf; number[2] 178; 215
piesterion *(πιεστήριον)*, press[1] 260
pinakas *(πίνακας)*, index 167
pinax *(πίναξ)*, plate; table 245; 333
pinax periechomenon *(πίναξ περιεχομένων)*, table
    of contents 334
pleres *(πλήρης)*, complete *a.* 85
pleres sylloge ton syngrammaton *(πλήρης συλλογή
    τῶν συγγραμμάτων)*, complete works 87
pleura prote *(πλευρά πρώτη)*, recto 277
poiema *(ποίημα)*, poem 248
poiesis *(ποίησις)*, poetry 250
poietes *(ποιητής)*, poet 249
polemike *(πολεμική)*, polemic 251
polyglossos *(πολύγλωσσος)*, polyglot 252
portraito *(πορτραῖτο)*, portrait 253
pragmateia *(πραγματεία)*, treatise 352
pragmateuomai *(πραγματεύομαι)*, treat 351
praktika *(πρακτικά)* pl., minutes *pl.* 202
praxis *(πρᾶξις)*, act 6
pro *(πρό)*, for 150
procheiros bibliotheke *(πρόχειρος βιβλιοθήκη)*,
    reference library 279

proedros *(πρόεδρος)*, president   259
proetoimazo *(προετοιμάζω)*, prepare   257
prokeryxe *(προκήρυξη)*, flysheet   148
prolegomena *(προλεγόμενα)*, introduction;
    preface   172; 256
prologos *(πρόλογος)*, introduction; preface   172; 256
prometopis *(προμετωπίς)*, title page   343
pros *(πρός)*, for; to   150; 344
prosago *(προσάγω)*, introduce   171
prosopografia *(προσωπογραφία)*, portrait   253
prosorine katalogographesis *(προσωρινή*
    καταλογογράφησις), temporary card   336
prosorinon deltion *(προσωρινόν δελτίον)*,
    temporary card   336
prosthekai *(προσθῆκαι) pl.*, addenda *pl.*   10
prostheke *(προσθήκη)*, appendix   26
prosthesis *(πρόσθεσις)*, appendix   26
prostithemi *(προστίθημι)*, annex   20
protos *(πρῶτος)*, first   147
prototypos *(πρωτότυπος)*, original   226
pseudonymon *(ψευδώνυμον)*, pseudonym   269

rahis *(ράχις)*, back   40
raster *(ράστερ)*, screen   298
retra *(ρήτρα)*, clause   70

satirikon periodikon *(σατιρικόν περιοδικόν)*,
    comic paper   79
schedion *(σχέδιον)*, sketch   312
schole *(σχολή)*, school   295
scholiazo *(σχολιάζω)*, comment[1]   81
scholion *(σχόλιον)*, commentary; explanation;
    note[1]   82; 141; 212
se *(σέ)*, in; on   164; 222
seira *(σειρά)*, series   305
selida *(σελίδα)*, page   228
selidosis *(σελίδωσις)*, paging   229
selis *(σελίς)*, page   228
semeiono *(σημειώνω)*, annotate   21
semeiosis *(σημείωσις)*, annotation   22
senarion *(σενάριον)*, scenario   294
skitso *(σκίτσο)*, sketch   312
spanios *(σπάνιος)*, rare   275
stele *(στήλη)*, column   77
sten *(στήν)*, → se *(σέ)*
stenographia *(στενογραφία)*, shorthand   308
stoicheiothesia *(στοιχειοθεσία)*, matter   195
strophe *(στροφή)*, strophe   325

syllego *(συλλέγω)*, collect   74
sylloge *(συλλογή)*, collection   75
sylloge tragudion *(συλλογή τραγουδιῶν)*, songbook   315
syllogike ekdosis *(συλλογική ἔκδοσις)*, corporate
    author   100
symbole *(συμβολή)*, contribution   93
sympleromatikon deltion *(συμπληρωματικόν*
    δελτίον), added entry   9
sympleromatikos *(συμπληρωματικός)*,
    supplementary   332
symplerono *(συμπληρώνω)*, complete *v.*   86
symposion *(συμπόσιον)*, assembly   32
syn *(σύν)*, with   368
syndeo *(συνδέω)*, sew   306
syndiaskepsis *(συνδιάσκεψις)*, conference   88
syndrome *(συνδρομή)*, subscription   328
synecheia *(συνέχεια)*, sequel   304
synechizo *(συνεχίζω)*, continue   92
synedriasis *(συνεδρίασις)*, sitting   311
synedrion *(αυνέδριον)*, congress   89
syneleusis *(συνέλευσις)*, assembly   32
synergates *(συνεργάτης)*, collaborator; joint
    author   73; 174
synergazomai *(συνεργάζομαι)*, collaborate   72
syngrammata mikta *(συγγράμματα μικτά) pl.*,
    miscellanea *pl.*   203
syngraphe *(συγγραφή)*, script; work   299; 371
syngrapheus *(συγγραφεύς)*, author; writer   35; 374
syngraphikon dikaioma *(συγγραφικόν δικαίωμα)*,
    copyright   98
syngrapho *(συγγράφω)*, elaborate; write   128; 372
synkentrotikos katalogos *(συγκεντρωτικός*
    κατάλογος), union catalogue   358
synopsizo *(συνοψίζω)*, summarize   329
syntaktes *(συντάκτης)*, editor[1]   125
syntasso *(συντάσσω)*, compile   84
syntasso katalogon *(συντάσσω κατάλογον)*,
    catalog *v.*   59
syntaxis *(σύνταξις)*, editorial office   127
syntomos anagraphe *(σύντομος ἀναγραφή)*,
    abbreviated entry   2
systello *(συστέλλω)*, abridge   4
systematike taxinomesis *(συστηματική*
    ταξινόμησις), classification   69

ta *(τά) pl.*, the   339
sten tainian magnetophonu *(στήν ταινίαν*
    μαγνητοφώνου), on tape   335

APPENDIX

CALENDAR UNITS — TEMPORA

Gamelion *(Γαμηλιών)*,
Ianuarios *('Ιανουάριος)*, January **I**
Anthesterion *('Ανθεστηριών)*,
Februarios *(Φεβρουάριος)*,
    February **II**
Elaphebolion *('Ελαφηβολιών)*,
Martios *(Μάρτιος)*, March **III**
Munichion *(Μουνιχιών)*,
Aprilios *('Απρίλιος)*, April **IV**
Thargelion *(Θαργηλιών)*,
Maios *(Μάϊος)*, May **V**
Skirophorion *(Σκιροφοριών)*,
Iunios *('Ιούνιος)*, June **VI**
Hekatombaion *('Εκατομβαιών)*,
Iulios *('Ιούλιος)*, July **VII**
Metageitnion *(Μεταγειτνιών)*,
Augustos *(Αὔγουστος)*, August **VIII**
Boedromion *(Βοηδρομιών)*,

Septembrios *(Σεπτέμβριος)*,
    September **IX**
Pyanopsion *(Πυανοψιών)*,
Oktobrios *('Οκτώβριος)*, October **X**
Maimakterion *(Μαιμακτηριών)*,
Noembrios *(Νοέμβριος)*,
    November **XI**
Poseideon *(Ποσειδεών)*,
Dekembrios *(Δεκέμβριος)*,
    December **XII**
Deutera *(Δευτέρα)*, Monday **XIII**
Trite *(Τρίτη)*, Tuesday **XIV**
Tetarte *(Τετάρτη)*, Wednesday **XV**
Pempte *(Πέμπτη)*, Thursday **XVI**
Paraskeue *(Παρασκευή)*,
    Friday **XVII**
Sabbaton *(Σάββατον)*,
    Saturday **XVIII**

Kyriake *(Κυριακή)*, Sunday **XXIX**
etos *(ἔτος)*, chronos, *(χρόνος)*,
    year **XX**
etesios *(ἐτήσιος)*, yearly **XXI**
hexamenos *ἐξάμενος* half-yearly
    semi-annual **XXII**
trimeniaios *(τριμηνιαῖος)*,
    quarterly **XXIII**
men *(μήν)*, month **XXIV**
meniaios *(μηνιαῖος)*, monthly **XXV**
dekapenthemeros *(δεκαπενθήμερος)*,
    fortnightly semi-monthly **XXVI**
hebdomas *(ἑβδομάς)*, week **XXVII**
hebdomadiaios *(ἑβδομαδιαῖος)*,
    weekly **XXVIII**
(he)mera *(ἡμέρα)*, day **XXIX**
hemeresios *(ἡμερήσιος)*,
    daily **XXX**

NUMERALS — NUMERI

1 **heis, mia, hen** *(εἷς, μία, ἕν)*   one
2 **dyo** *(δύο)*   two
3 **treis, tria** *(τρεῖς, τρία)*   three
4 **tettares, tettara** *(τέτταρες, -α)*   four
5 **pente** *(πέντε)*   five
6 **hex, hexe** *(ἕξ, ἕξη)*   six
7 **hepta** *(ἑπτά)*   seven
8 **okto** *(ὀκτώ)*   eight

9 **ennea** *(ἐννέα)*   nine
10 **deka** *(δέκα)*   ten
11 **hendeka** *(ἕνδεκα)*   eleven
12 **dodeka** *(δώδεκα)*   twelve
13 **treis/tria kai deka**
    τρεῖς/τρία καί δέκα   thirteen
14 **tettares/tettara kai deka**
    (τέτταρες/τέτταρα καί δέκα)   fourteen

15 pentekaideka *(πεντεκαίδεκα)*	fifteen
16 hekkaideka *(ἑκκαίδεκα)*	sixteen
17 heptakaideka *(ἑπτακαίδεκα)*	seventeen
18 oktokaideka *(ὀκτωκαίδεκα)*	eighteen
19 enneakaideka *(ἐννεακαίδεκα)*	nineteen
20 eikosi(n) *(εἴκοσιν)*	twenty
21 heis kai eikosi *(εἷς καί εἴκοσι)*	twenty one
30 triakonta *(τριάκοντα)*	thirty
40 tettarakonta *(τετταράκοντα)*	forty
50 pentekonta *(πεντήκοντα)*	fifty
60 hexekonta *(ἑξήκοντα)*	sixty
70 hebdomekonta *(ἑβδομήκοντα)*	seventy
80 ogdoekonta *(ὀγδοήκοντα)*	eighty
90 enenekonta *(ἐνενήκοντα)*	ninety
100 hekaton *(ἑκατόν)*	hundred
200 diakosioi *(διακόσιοι)*	two hundred
300 triakosioi *(τριακόσιοι)*	three hundred
400 tetrakosioi *(τετρακόσιοι)*	four hundred
500 pentakosioi *(πεντακόσιοι)*	five hundred
600 hexakosioi *(ἑξακόσιοι)*	six hundred
700 heptakosioi *(ἑπτακόσιοι)*	seven hundred
800 oktakosioi *(ὀκτακόσιοι)*	eight hundred
900 enakosioi *(ἐνακόσιοι)*	nine hundred
1000 chilioi *(χίλιοι)*	thousand
1. protos, -e, -on *(πρῶτος, -η, -ον)*	first
2. deuteros *(δεύτερος)*	second
3. tritos *(τρίτος)*	third
4. tetartos *(τέταρτος)*	fourth
5. pemptos *(πέμπτος)*	fifth
6. hektos *(ἕκτος)*	sixth
7. hebdomos *(ἕβδομος)*	seventh
8. ogdoos *(ὄγδοος)*	eighth
9. enatos *(ἔνατος)*	ninth
10. dekatos *(δέκατος)*	tenth
11. hendekatos *(ἑνδέκατος)*	eleventh

12. dodekatos *(δωδέκατος)*	twelfth
13. tritos kai dekatos *(τρίτος καί δέκατος)*	thirteenth
14. tetartos kai dekatos *(τέταρτος καί δέκατος)*	fourteenth
15. pemptos kai dekatos *(πέμπτος καί δέκατος)*	fifteenth
16. hektos kai dekatos *(ἕκτος καί δέκατος)*	sixteenth
17. hebdomos kai dekatos *(ἕβδομος καί δέκατος)*	seventeenth
18. ogdoos kai dekatos *(ὄγδοος καί δέκατος)*	eighteenth
19. enatos kai dekatos *(ἔνατος καί δέκατος)*	nineteenth
20. eikostos *(εἰκοστός)*	twentieth
21. heis kai eikostos *(εἷς καί εἰκοστός)*	twenty first
30. triakostos *(τριακοστός)*	thirtieth
40. tettarakostos *(τετταρακοστός)*	fortieth
50. pentekostos *(πεντηκοστός)*	fiftieth
60. hexekostos *(ἑξηκοστός)*	sixtieth
70. hebdomekostos *(ἑβδομηκοστός)*	seventieth
80. ogdoekostos *(ὀγδοηκοστός)*	eightieth
90. enenekostos *(ἐνενηκοστός)*	ninetieth
100. hekatostos *(ἑκατοστός)*	hundredth
200. diakosiostos *(διακοσιοστός)*	two hundredth
300. triakosiostos *(τριακοσιοστός)*	three hundredth
400. tetrakosiostos *(τετρακοσιοστός)*	four hundredth
500. pentakosiostos *(πεντακοσιοστός)*	five hundredth
600. hexakosiostos *(ἑξακοσιοστός)*	six nundredth
700. heptakosiostos *(ἑπτακοσιοστός)*	seven hundredth
800. oktakosiostos *(ὀκτακοσιοστός)*	eight hundredth
900. enakosiostos *(ἐνακοσιοστός)*	nine hundredth
1000. chiliostos *(χιλιοστός)*	thousandth

GREEK ALPHABET. TRANSLITERATION

LITTERAE GRAECAE. TRANSLITTERATIO

A — α, **a**	*Θ — θ,* **th**	*Σ — σ, ς,* **s**
B — β, **b**	*I — ι,* **i**	*T — τ,* **t**
Γ — γ, **g**	*K — κ,* **k**	*Y — υ,* **y**
(γγ, **ng**	*Λ — λ,* **l**	*(αυ,* **au**
γκ, **nk**	*M — μ,* **m**	*ευ,* **eu**
γχ, **nch)**	*N — ν,* **n**	*ου,* **u)**
Δ — δ, **d**	*Ξ — ξ,* **x**	*Φ — φ,* **ph**
E — ε, **e**	*O — ο,* **o**	*X — χ,* **ch**
Z — ζ, **z**	*Π — π,* **p**	*Ψ — ψ,* **ps**
H — η, **e**	*P — ϱ,* **r**	*Ω — ω,* **o**

HUNGARIAN—HUNGARICA

APPENDIX

CALENDAR UNITS — TEMPORA

január, January	I	**december,** December	XII	**félévi,** half-yearly, semi-annual	XXII		
február, February	II	**hétfő,** Monday	XIII	**negyedévi,** quarterly	XXIII		
március, March	III	**kedd,** Tuesday	XIV	**hó, hónap,** month	XXIV		
április, April	IV	**szerda,** Wednesday	XV	**havi,** monthly	XXV		
május, May	V	**csütörtök,** Thursday	XVI	**félhavi,** fortnightly, semi-monthly	XXVI		
június, June	VI	**péntek,** Friday	XVII	**hét,** week	XXVII		
július, July	VII	**szombat,** Saturday	XVIII	**heti,** weekly	XXVIII		
augusztus, August	VIII	**vasárnap,** Sunday	XIX	**nap,** day	XXIX		
szeptember, September	IX	**év,** year	XX	**napi,** daily	XXX		
október, October	X	**évi,** yearly, annual	XXI				
november, November	XI						

NUMERALS — NUMERI

1	**egy**	one		16	**tizenhat**	sixteen	
2	**kettő**	two		17	**tizenhét**	seventeen	
3	**három**	three		18	**tizennyolc**	eighteen	
4	**négy**	four		19	**tizenkilenc**	nineteen	
5	**öt**	five		20	**húsz**	twenty	
6	**hat**	six		21	**huszonegy**	twenty one	
7	**hét**	seven		30	**harminc**	thirty	
8	**nyolc**	eight		40	**negyven**	forty	
9	**kilenc**	nine		50	**ötven**	fifty	
10	**tíz**	ten		60	**hatvan**	sixty	
11	**tizenegy**	eleven		70	**hetven**	seventy	
12	**tizenkettő**	twelve		80	**nyolcvan**	eighty	
13	**tizenhárom**	thirteen		90	**kilencven**	ninety	
14	**tizennégy**	fourteen		100	**száz**	hundred	
15	**tizenöt**	fifteen		200	**kétszáz**	two hundred	

300 háromszáz	three hundred		16. tizenhatodik	sixteenth
400 négyszáz	four hundred		17. tizenhetedik	seventeenth
500 ötszáz	five hundred		18. tizennyolcadik	eighteenth
600 hatszáz	six hundred		19. tizenkilencedik	nineteenth
700 hétszáz	seven hundred		20. huszadik	twentieth
800 nyolcszáz	eight hundred		21. huszonegyedik	twenty first
900 kilencszáz	nine hundred		30. harmincadik	thirtieth
1000 ezer	thousand		40. negyvenedik	fortieth
1. első	first		50. ötvenedik	fiftieth
2. második	second		60. hatvanadik	sixtieth
3. harmadik	third		70. hetvenedik	seventieth
4. negyedik	fourth		80. nyolcvanadik	eightieth
5. ötödik	fifth		90. kilencvenedik	ninetieth
6. hatodik	sixth		100. századik	hundredth
7. hetedik	seventh		200. kétszázadik	two hundredth
8. nyolcadik	eighth		300. háromszázadik	three hundredth
9. kilencedik	ninth		400. négyszázadik	four hundredth
10. tizedik	tenth		500. ötszázadik	five hundredth
11. tizenegyedik	eleventh		600. hatszázadik	six hundredth
12. tizenkettedik	twelfth		700. hétszázadik	seven hundredth
13. tizenharmadik	thirteenth		800. nyolcszázadik	eight hundredth
14. tizennegyedik	fourteenth		900. kilencszázadik	nine hundredth
15. tizenötödik	fifteenth		1000. ezredik	thousandth

HUNGARIAN ALPHABET

LITTERAE HUNGARICAE

A a, Á á	Gy gy	Ny ny	Ty ty
B b	H h	O o, Ó ó	U u, Ú ú
C c	I i, Í í	Ö ö, Ő ő	Ü ü, Ű ű
Cs cs	J j	P p	V v
D d	K k	Q q	W w
Dz dz	L l	R r	X x
Dzs dzs	Ly ly	S s	Y y
E e, É é	M m	Sz sz	Z z
F f	N n	T t	Zs zs
G g			

ITALIAN – ITALICA

duplicato, copy[1]; duplicate	96; 121
e, and	18
ed, and	18
editore, editor[2]; publisher[1]	126; 273
edizione, edition	124
edizione originale, original edition	227
edizione principe, original edition	227
elaborare, elaborate	128
elegia, elegy	129
elenco, list	184
elogio, panegyric	230
elucubrare, elaborate	128
emendare, correct	101
enciclopedia, encyclopaedia	130
ente collettivo, corporate author	100
epilogo, epilogue	135
epitalamio, bridal song	53
epitome, extract	142
epopea, epic	134
errore di stampa, printer's error	265
esemplare, copy[2]	97
esemplare d'obbligo, deposit copy	112
esporre, explain	140
esposizione, exhibition	139
estratto, extract; offprint	142; 221
facsimile, facsimile	144
falsificazione, forgery	151
fascicolo, number[2]	215
fiaba, fable	143
figura, picture	241
foglio, leaf; sheet	178; 307
foglio volante, flysheet	148
fonte, source	316
fotografia, photograph	240
fototipia, collotype	76
frammento, fragment	152
frontespizio, title page	343
gazzetta, newspaper	209
giornale, diary; newspaper	115; 209
giornale umoristico, comic paper	79
gli pl., the	339
glossario, glossary	154

grande, large	176
guida, guide(-book)	155
i pl., the	339
iconografia, iconography	158
idillio, idyll	159
il, the	339
illuminare, explain	140
illustrare, illustrate; represent	160; 284
illustrazione, commentary; picture	82; 241
imitazione, imitation	161
immagine, picture	241
impiego, use n.	362
imprimere, print[1]	261
in, in	164
incerto, doubtful	119
incidere, engrave	131
incisione, engraving	132
incisione all'acquaforte, etching	137
incisione in legno, woodcut	370
incisione in rame, copperplate engraving	95
incompleto, incomplete	165
incunabulo, incunabula pl.	166
indicatore, index	167
indicazione abbreviata, abbreviated entry	2
indice, index	167
indice delle materie, table of contents	334
inedito, unpublished	361
iniziale, initial	168
inno, anthem; panegyric	24; 230
inserire, insert	169
intercalare, insert	169
intero, complete	85
interpretare, translate	348
introdurre, introduce	171
introduzione, introduction	172
invariato, unchanged	356
inventario, inventory	173
istituto, institute	170
l', the	339
la, the	339
lascito, literary remains pl.	185
le pl., the	339
legare, bind	45
legare (alla rustica), sew	306

presidente, president 259
pressa, press[1] 260
prestito, lending 180
primario, chief 66
primo, chief; first 66; 147
processo verbale, minutes *pl.* 202
prosa, prose 267
protocollo, minutes *pl.* 202
provvedere, provide (with) 268
pseudonimo, pseudonym 269
pubblicare, edit[2]; publish 123; 271
essere **pubblicato**, *be* published 272
pubblicazione, edition; transactions *pl.* 124; 346
pubblicazione cessata, ceased publication 61
pubblicazione commemorativa, memorial volume 199
pubblicità, advertisement 11

quaderno, booklet 50

raccogliere, collect 74
raccolta, collection 75
racconto, story 324
raffigurare, represent 284
rapporto, report[1] 283
rappresentare, represent 284
raro, rare 275
rassegna letteraria, review *n.* 287
recensione, critique 104
recensire, review *v.* 288
recente, new 208
recto, recto 277
redattore, editor[1] 125
redazione, editorial office 127
registro, list 184
regola(mento), prescription; statute[1] 258; 323
relazione, report[1] 283
repertorio, repertory 282
reticolo, screen 298
revisione, revision 290
riassumere, abridge; summarize 4; 329
ricordi *pl.*, memoirs *pl.* 197
rielaborare, rewrite 291
rifondere, rewrite 291
rilegare, bind 45
rilegatura, binding 47
rilegatura alla rustica, paperback 232

rinnovato, new 208
riproduzione, reproduction 286
ristampa, reprint[1] 285
ritoccare, rewrite 291
ritratto, portrait 253
rivedere, revise[1] 289
rivista, periodical[2]; review *n.* 239; 287
rivista di moda, fashion magazine 145
romanzi e novelle *pl.*, fiction 146
romanzo, novel 213
romanzo di fantascienza, science fiction 297

saggio, essay[1] 136
scala, scale 293
scegliere, select 302
scenario, scenario 294
scheda, card 56
scheda di rinvio, reference card 278
scheda principale, main card 191
scheda provvisoria, temporary card 336
schedario, card index 57
scheda secondaria, added entry 9
schizzo, sketch 312
scienza, science 296
scompleto, incomplete 165
scritto, script 299
scrittore, writer 374
scrittura, script 299
scrittura Braille, braille printing 52
scrivere, write 372
scuola, school 295
secolo, century 62
segnatura, location mark; signature 189; 310
sentenza, adage 7
senza, without 369
serie, series 305
servizio degli scambi, exchange centre 138
sessione, sitting 311
sezione, section 301
s. l. s. d. → *senza* luogo senza data
società, society 313
sommario, summary; table of contents 330; 334
sopra, about; on 3; 222
soscrizione, imprint 163
sotto, under 357
spento (relativo a periodici), ceased publication 61
spiegare, explain 140

APPENDIX

CALENDAR UNITS — TEMPORA

gennaio, January	I	dicembre, December	XII	semestrale, half-yearly,	
febbraio, February	II	lunedì, Monday	XIII	semi-annual	XXII
marzo, March	III	martedì, Tuesday	XIV	trimestrale, quarterly	XXIII
aprile, April	IV	mercoledì, Wednesday	XV	mese, month	XXIV
maggio, May	V	giovedì, Thursday	XVI	mensile, monthly	XXV
giugno, June	VI	venerdì, Friday	XVII	semimensile, fortnightly,	
luglio, July	VII	sabato, Saturday	XVIII	semi-monthly	XXVI
agosto, August	VIII	domenica, Sunday	XIX	settimana, week	XXVII
settembre, September	IX	anno, year	XX	settimanale, weekly	XXVIII
ottobre, October	X	annuale, yearly	XXI	giorno, day	XXIX
novembre, November	XI			quotidiano, daily	XXX

NUMERALS — NUMERI

1	uno, una, un, un'	one	16	sedici	sixteen
2	due	two	17	diciassette	seventeen
3	tre	three	18	diciotto	eighteen
4	quattro	four	19	diciannove	nineteen
5	cinque	five	20	venti	twenty
6	sei	six	21	ventuno	twenty one
7	sette	seven	30	trenta	thirty
8	otto	eight	40	quaranta	forty
9	nove	nine	50	cinquanta	fifty
10	dieci	ten	60	sessanta	sixty
11	undici	eleven	70	settanta	seventy
12	dodici	twelve	80	ottanta	eighty
13	tredici	thirteen	90	novanta	ninety
14	quattordici	fourteen	100	cento	hundred
15	quindici	fifteen	200	duecento	two hundred

300	trecento	three hundred
400	quattrocento	four hundred
500	cinquecento	five hundred
600	seicento	six hundred
700	settecento	seven hundred
800	ottocento	eight hundred
900	novecento	nine hundred
1000	mille	thousand
1.	primo, -a	first
2.	secondo	second
3.	terzo	third
4.	quarto	fourth
5.	quinto	fifth
6.	sesto	sixth
7.	settimo	seventh
8.	ottavo	eighth
9.	nono	ninth
10.	decimo	tenth
11.	undicesimo, decimoprimo	eleventh
12.	dodicesimo, decimosecondo	twelfth
13.	tredicesimo, decimoterzo	thirteenth
14.	quattordicesimo, decimoquarto	fourteenth
15.	quindicesimo, decimoquinto	fifteenth

16.	sedicesimo, decimosesto	sixteenth
17.	diciassettesimo, decimosettimo	seventeenth
18.	diciottesimo, decimottavo	eighteenth
19.	diciannovesimo, decimonono	nineteenth
20.	ventesimo	twentieth
21.	ventunesimo, ventesimo primo	twenty-first
30.	trentesimo	thirtieth
40.	quarantesimo	fortieth
50.	cinquantesimo	fiftieth
60.	sessantesimo	sixtieth
70.	settantesimo	seventieth
80.	ottantesimo	eightieth
90.	novantesimo	ninetieth
100.	centesimo	hundredth
200.	du(e)centesimo	two hundredth
300.	trecentesimo	three hundredth
400.	quattrocentesimo	four hundredth
500.	cinquecentesimo	five hundredth
600.	seicentesimo	six hundredth
700.	settecentesimo	seven hundredth
800.	ottocentesimo	eight hundredth
900.	novecentesimo	nine hundredth
1000.	millesimo	thousandth

LATIN—LATINA

praescriptum, order[1]; prescription; statute[1] 224; 258; 323
praeses, president 259
prelum typographicum, press[1] 260
primus, chief; first 66; 147
princeps, chief; first 66; 147
pro, for 150
procœmium, preface 256
propaganda, advertisement 11
propter, for 150
prosa, prose 267
prosa pulchra, fiction 146
prosa scientifica phantastica, science fiction 297
protocollum, minutes *pl.* 202
pseudonymus, pseudonym 269
publicare, edit[2], publish 123; 271
publicari, *be* published 272
publicatio cessata, ceased publication 61
publicus, official 220

rarus, rare 275
raster, screen 298
recens, new 208
recensere, criticize; review 105; 288
recensio, critique 104
recognitio, revision 290
recognoscere, revise[1] 289
recto, recto 277
redactio, editorial office 127
redactor, editor[1] 125
redemptor, editor 273
refer(a)tum, report[1] 283
refutatio, refutation 281
regestrum, minutes *pl.* 202
register, list 184
regula, prescription; statute[1] 258; 323
reimpressum, reprint[1] 285
repertorium, repertory 282
reproductio, reproduction 286
res gestae *pl.*, history 157
retractare, rewrite 291
revidere, revise[1] 289
revisio, revision 290
rudimentum, textbook 338

saeculum, century 62
scala, scale 293

scenarium, scenario 294
schola, school 295
scida, card 56
scida adiuncta, added entry 9
scida catalogi, card 56
scida principalis, main card 191
scida referentiae, reference card 278
scida secunda, added entry 9
scida temporaria, temporary card 336
scientia, science 296
scribere, write 372
scriptor, author; writer 35; 374
scriptum, document; script 117; 299
scriptura, script 299
scriptura Braille (pro caecis), braille printing 52
sculpere, engrave 131
sculptura, engraving 132
sectio, section 301
secundum, after 12
seligere, select 302
sententia, adage 7
separatum, offprint 221
series, series 305
series illustrationum delectantium in diurnis, comic strip 80
sermo, dialog, speech 114; 319
sessio, sitting 311
signatura, signature 310
signatura loci, location mark 189
sine, without 369
s. l. s. a. → *sine* loco sine anno
societas, society 313
solus, only *a.* 223
stenographia, shorthand 308
stropha, strophe 325
studium, essay[1]; study 136; 326
sub, under 357
subter, under 357
summa, contents *pl.* 91
super, about; on 3; 222
supplementum, addenda; supplementary 10, 332
supplere, complete 86
suppletivus, supplementary 332
supra, about; on 3; 222

tabella, table 333
taberna libraria, bookshop 51

tabula, plate	245
tabulae *pl.,* list; minutes *pl.*	184; 202
tachygraphia, shorthand	308
taenia pellicularis vocis, on tape	335
taeniola photographica, microfilm	200
tegumentum/tegimentum libri, binding	47
textus, text	337
thesaurus, collection	75
thesaurus verborum, dictionary	116
titulus, title	341
titulus currens, running title	292
tomus, volume[1]	366
totus, complete	85
tractare, treat	351
tractatio, treatise	352
tractatulus, treatise	352
tragœdia, tragedy	345
transcriptio, transcription	347
transferre, translate	348
translatio, translation	349
translitteratio, transliteration	350
typographia, printing house; typography	266; 355
typographus, printer	264
typus, impression	162
una → **unus**	
unicus, only *a.*	223

universitas (scientiarum), university	359
unus, una, unum, a, an; only *a.*	1; 223
usus, use *n.*	362
uti, use *v.*	363
vademecum, pocket-book	247
varians, version	364
variare, change[1]	63
variari, change[2]	64
variatio, version	364
verbum, adage	7
verso, verso	365
versus, poem	248
versus facere, write poetry	373
vertere, translate	348
verus, authentic	34
vetus, ancient	17
vocabularium, dictionary; glossary	116; 154
vocabulum, subject heading	327
vocabulum obiecti, title entry	342
vocabulum rei, title entry	342
vocabulum tituli, title entry	342
volumen, volume[1]	366
vox, subject heading	327
xerographia, woodcut; xerography	370; 375

APPENDIX

CALENDAR UNITS — TEMPORA

Januarius, January	I	**Dies Martis, feria III.,**		**semestris,** half-yearly,	
Februarius, February	II	Tuesday	XIV	semi-annual	**XXII**
Martius, March	III	**Dies Mercurii, feria IV.,**		**trimestris,** quarterly	**XXIII**
Aprilis, April	IV	Wednesday	XV	**mensis,** month	**XXIV**
Maius, May	V	**Dies Iovis, feria V.,**		**mensilis, menstruus,**	
Junius, June	VI	Thursday	XVI	monthly	**XXV**
Julius, July	VII	**Dies Veneris, feria VI.,**		**semimensilis, semimen-**	
Augustus, August	VIII	Friday	XVII	**struus,** fortnightly, semi-	
September, September	IX	**Dies Saturni, feria VII.,**		monthly	**XXVI**
October, October	X	Saturday	XVIII	**hebdomas, hebdomada,** week	**XXVII**
November, November	XI	**Dies Solis, Dominica,**		**semel in hebdomade/hebdo-**	
December, December	XII	Sunday	XIX	**mada,** weekly	**XXVIII**
Dies Lunae, feria II.,		**annus,** year	XX	**dies,** day	**XXIX**
Monday	XIII	**annalis, annuus,** annual,		**cotidianus,** daily	**XXX**
		yearly	XXI		

NUMERALS — NUMERI

1 (I)	**unus, -a, -um**	one	13 (XIII)	**tredecim**	thirteen
2 (II)	**duo, duae, duo**	two	14 (XIV)	**quattuordecim**	fourteen
3 (III)	**tres, tria**	three	15 (XV)	**quindecim**	fifteen
4 (IV)	**quattuor**	four	16 (XVI)	**sedecim**	sixteen
5 (V)	**quinque**	five	17 (XVII)	**septendecim**	seventeen
6 (VI)	**sex**	six	18 (XVIII)	**duodeviginti, octodecim**	eighteen
7 (VII)	**septem**	seven	19 (XIX)	**undeviginti, novemdecim**	nineteen
8 (VIII)	**octo**	eight	20 (XX)	**viginti**	twenty
9 (IX)	**novem**	nine	21 (XXI)	**unus et viginti**	twenty one
10 (X)	**decem**	ten	30 (XXX)	**triginta**	thirty
11 (XI)	**undecim**	eleven	40 (XL)	**quadraginta**	forty
12 (XII)	**duodecim**	twelve	50 (L)	**quinquaginta**	fifty

60 (LX)	sexaginta	sixty	
70 (LXX)	septuaginta	seventy	
80 (LXXX)	octoginta	eighty	
90 (XC)	nonaginta	ninety	
100 (C)	centum	hundred	
200 (CC)	ducenti, -ae, -a	two hundred	
300 (CCC)	trecenti	three hundred	
400 (CD)	quadringenti	four hundred	
500 (D)	quingenti	five hundred	
600 (DC)	sescenti	six hundred	
700 (DCC)	septingenti	seven hundred	
800 (DCCC)	octingenti	eight hundred	
900 (CM)	nongenti	nine hundred	
1000 (M)	mille	thousand	

1. primus, -a, -um — first
2. secundus, alter — second
3. tertius — **third**
4. quartus — fourth
5. quintus — fifth
6. sextus — sixth
7. septimus — seventh
8. octavus — eighth
9. nonus — ninth
10. decimus — tenth
11. undecimus — eleventh
12. duodecimus — twelfth

13. tertius decimus — thirteenth
14. quartus decimus — fourteenth
15. quintus decimus — fifteenth
16. sextus decimus — sixteenth
17. septimus decimus — seventeenth
18. duodevicesimus — eighteenth
19. undevicesimus — nineteenth
20. vicesimus — twentieth
21. vicesimus primus — twenty first
30. tricesimus — thirtieth
40. quadragesimus — fortieth
50. quinquagesimus — fiftieth
60. sexagesimus — sixtieth
70. septuagesimus — seventieth
80. octogesimus — eightieth
90. nonagesimus — ninetieth
100. centesimus — hundredth
200. ducentesimus — two hundredth
300. trecentesimus — three hundredth
400. quadringentesimus — four hundredth
500. quingentesimus — five hundredth
600. sescentesimus — six hundredth
700. septingentesimus — seven hundredth
800. octingentesimus — eight hundredth
900. nongentesimus — nine hundredth
1000. millesimus — thousandth

NORWEGIAN — NORVEGICA

foreløpig seddel, temporary card — 336
forfalskning, forgery — 151
forfatter, author; writer — 35; 374
forfatternavn, pseudonym — 269
forfatterrett, copyright — 98
forhandlinger, report[1] — 283
forklare, explain — 140
forkorte, abridge — 4
forkortet innførsel, abbreviated entry — 2
forlag, publishing house — 274
forlegge, edit[2]; publish — 123; 271
forlegger, publisher[1] — 273
formann, president — 259
forme, represent — 284
forord, preface — 256
forordning, order[1] — 224
forsamling, assembly — 32
forside, recto — 277
på forskjellige steder pl., passim — 237
forskrift, prescription; statute[1] — 258; 323
forsvare, defend — 111
forsyne med ..., provide (with) — 268
forsyne med anmerkninger, annotate — 21
fortegnelse, list — 184
fortelling, short story; story — 309; 324
fortsette, continue — 92
fortsettelse, sequel — 304
fotnote, footnote — 149
fotografi, photograph — 240
fra, from; of — 153; 218
fragment, fragment — 152
framstille, represent — 284
fremtidsromaner pl., science fiction — 297
fullstendiggjøre, complete v. — 86
følge, series — 305
fører, guide(-book) — 155
første, first — 147
første utgave, original edition — 227

gammel, ancient — 17
gjendrivelse, refutation — 281
gjennomse, revise[1] — 289
gjenoppta, summarize — 329
gjenpart, duplicate — 121
glossar, glossary — 154
grammofonplate, record[1] — 276
gravere, engrave — 131

gravering, engraving — 132
på grunn av ..., based on the ... — 41
grunnriss, sketch — 312
guide, guide(-book) — 155
gått inn, ceased publication — 61

hefte n. booklet; number[2] — 50; 215
hefte v. sew — 306
heftet (bok), paperback — 232
heltedikt, epic — 134
henvisningskort, reference card — 278
historie, history — 157
hoved, chief — 66
hovedkort, main card — 191
hovedseddel, main card — 191
hymne, anthem — 24
hyrdedikt, idyll — 159
håndbibliotek, reference library — 279
håndbok, handbook; pocket-book — 156; 247
håndskript, manuscript — 192

i, in — 164
idyll, idyll — 159
ikke offentliggjort, unpublished — 361
ikke udgivet, unpublished — 361
ikonografi, iconography — 158
illustrasjon, picture — 241
illustrere, illustrate — 160
imitasjon, imitation — 161
impressum, imprint — 163
index, index — 167
initiale, initial — 168
inkunabel, incunabula pl. — 166
innberetning, report[1] — 283
innbinde, bind — 45
innbinding, binding — 47
inneholde, contain — 90
innføre, insert; introduce — 169; 171
innførelse, introduction — 172
innføring, introduction — 172
inngravere, engrave — 131
innhold, contents; table of contents — 91; 334
innholdsfortegnelse, table of contents — 334
innlede, introduce — 171
innledning, introduction — 172
innsette, insert — 169

samling, collection	75
sammendrag, summary	300
sammenfatte, summarize	329
sammenstille, compile	84
samtale, dialog	114
sang, song	314
sangbok, songbook	315
sats, matter	195
science fiction, science fiction	297
seddelkatalog, card index	57
selskap, society	313
selvbiografi, autobiography	38
sentens, adage	7
separatavtrykk, offprint	221
serie, series	305
sesjon, sitting	311
sette til, annex	20
side, page	228
signatur, location mark; signature	189; 310
sjef, chief	66
sjelden, rare	275
skisse, sketch	312
skjære, engrave	131
skole, school	295
skolebok, textbook	338
skrift, script	299
skrifter pl., transactions pl.	346
skriftgrad, type size	354
skrive, write	372
skrivning, script	299
skuespill, play	246
slagord, subject heading	327
slutningsbemerkning, epilogue	135
små pl., little	188
spalte, column	77
språk, language; speech	175; 319
standardisering, standardisation	321
statsbibliotek, state library	322
sted, place	243
uten sted uten år (u. s. u. å.), no place no date	210
steintrykk, lithograph(y)	187
stenografi, shorthand	308
stor, large	176
strofe, strophe	325
studium, study	326
stykke, piece	242
subskripsjon, subscription	328
supplement, addenda	10

supplere, complete	86
systematisering, classification	69
sørgespill, tragedy	345
tabell, table	333
tale, speech	319
tall, number[1]	214
tavle, plate; table	245; 333
tegneserie, comic strip	80
tegning, drawing	120
tekst, text	337
tidsskrift, periodical[2]	239
til, to	344
tilegnelse, dedication	109
tillegg, addenda; appendix; supplement n.	10; 26; 331
tilleggs-, supplementary	332
tilleggsinnførsel, added entry	9
tillempe, adapt	8
tittel, title	341
tittelblad, title page	343
tospråklig, bilingual	44
tragedie, tragedy	345
traktere, treat	351
translitterasjon, transliteration	350
transskripsjon, transcription	347
tresnitt, woodcut	370
trykk, impression; press[1]	162; 260
trykke, print[1]	261
trykkeri, printing house	266
trykkfeil, printer's error	265
trykksaker, printed matter	262
trykkår, date of printing	106
tvilsom, doubtful	119
typografi, typography	355
uforanderlig, unchanged	356
uforandret, unchanged	356
ufulstendig, incomplete	165
uinnbundet (bok), paperback	232
ukomplett, incomplete	165
under, under	357
universitet, university	359
universitetsbibliotek, university library	360
u. s. u. å. → uten sted uten år	
utdrag, extract	142

APPENDIX

CALENDAR UNITS – TEMPORA

januar, January	I	desember, December	XII	halvårlig, half-yearly,	
februar, February	II	mandag, Monday	XIII	semi-annual	XXII
mars, March	III	tirsdag, Tuesday	XIV	kvartal, quarterly	XXIII
april, April	IV	onsdag, Wednesday	XV	måned, month	XXIV
mai, May	V	torsdag, Thursday	XVI	månedlig, monthly	XXV
juni, June	VI	fredag, Friday	XVII	halvmånedlig, toukentlig, fort-	
juli, July	VII	lørdag, Saturday	XVIII	nightly, semi-monthly	XXVI
august, August	VIII	søndag, Sunday	XIX	uke, week	XXVII
september, September	IX	år, year	XX	ukentlig, weekly	XXVIII
oktober, October	X	årlig, annual, yearly	XXI	dag, day	XXIX
november, November	XI			daglig, daily	XXX

NUMERALS – NUMERI

1	en, ett	one	16 seksten	sixteen
2	to	two	17 sytten	seventeen
3	tre	three	18 atten	eighteen
4	fire	four	19 nitten	nineteen
5	fem	five	20 tyve	twenty
6	seks	six	21 en og tyve	twenty-one
7	syv	seven	30 tredve	thirty
8	åtte	eight	40 førti	forty
9	ni	nine	50 femti	fifty
10	ti	ten	60 seksti	sixty
11	elleve	eleven	70 sytti	seventy
12	tolv	twelve	80 åtti	eighty
13	tretten	thirteen	90 nitti	ninety
14	fjorten	fourteen	100 hundre	hundred
15	femten	fifteen	200 to hundre	two hundred

300 tre hundre	three hundred	17. syttende	seventeenth
400 fire hundre	four hundred	18. attende	eighteenth
500 fem hundre	five hundred	19. nittende	nineteenth
600 seks hundre	six hundred	20. tyvende	twentieth
700 syv hundre	seven hundred	21. enogtivende	twenty-first
800 åtte hundre	eight hundred	30. tredevte	
900 ni hundre	nine hundred	tredje?trettiende	thirtieth
1000 tusen	thousand	40. førtiende	fortieth
1. første	first	50. femtiende	fiftieth
2. annen, annet	second	60. sekstiende	sixtieth
3. tredje	third	70. syttiende	seventieth
4. fjerde	fourth	80. åttiende	eightieth
5. femte	fifth	90. nittiende	ninetieth
6. sjette	sixth	100. hundrede	hundredth
7. syvende	seventh	200. to hundrede	two hundredth
8. åttende	eighth	300. tre hundrede	three hundredth
9. niende	ninth	400. fire hundrede	four hundredth
10. tiende	tenth	500. fem hundrede	five hundredth
11. ellevte	eleventh	600. seks hundrede	six hundredth
12. tolvte	twelfth	700. syv hundrede	seven hundredth
13. trettende	thirteenth	800. åtte hundrede	eight hundredth
14. fjortende	fourteenth	900. ni hundrede	nine hundredth
15. femtende	fifteenth	1000. tusende	thousandth
16. sekstende	sixteenth		

NORWEGIAN ALPHABET — LITTERAE NORVEGICAE

A a	I i	R r	Æ æ
B b	J j	S s	Ø ø
C c	K k	T t	Å å
D d	L l	U u	
E e	M m	V v	
F f	N n	W w	
G g	O o	X x	
H h	P p	Y y	
	Q q	Z z	

POLISH — POLONICA

22

POLISH — POLONICA

uniwersytet, university 359
upoważnić, authorize 37
urywek, fragment 152
urząd, office 219
urzędowy, official 220
ustawa, law; statute[1] 177; 323
ustęp, paragraph 234
utwór, work 371
utwór dramatyczny, play 246
uwaga, footnote; note[1] 149; 212
uzupełniać, complete v. 86
uzupełnienie, appendix 26
użycie, use n. 362
użytek, use n. 362
użytkować, use v. 363

w, in 164
wariant, version 364
wątpliwy, doubtful 119
we, in 164
wersja, version 364
wiadomości pl., transactions pl. 346
wiązać, bind 45
wiedza, science 296
wiek, century 62
wielki, large 176
wielojęzyczny, polyglot 252
wiersz, poem 248
wprowadzić, introduce 171
współautor, joint author 174
współpracować, collaborate 72
współpracownik, collaborator 73
wspomnienie, memoirs pl. 197
wstęp, introduction; preface 172; 256
wszystek, complete a 85
wybierać, select 302
wyciąg, extract 142
wydanie, edition 124
wydanie kompletne, complete works 87
wydawać, edit[2]; publish 123; 271
wydawca, editor[2]; publisher[1] 126; 273
wydawnictwo, publishing house 274
wydawnictwo informacyjne, reference work 280
wydawnictwo jubileuszowe, memorial volume 199

wydawnictwo luźnokartkowe, loose-leaf book 190
wydawnictwo nutowe, printed music 263
wydawnictwo skoroszytowe, loose-leaf book 190
wydawnictwo zawieszone, ceased publication 61
wyjaśniać, explain 140
wyjaśnienie, commentary 82
wykład, lecture 179
wypisy, selection 303
wypożyczanie, lending 180
wystawa, exhibition 139

z, from; of; on; with 153; 218; 222; 368
za, after; for 12; 150
załącznik, supplement n. 331
załączyć, annex 20
załatwić, edit[1] 122
zamówienie, order[2] 225
zarys, sketch 312
zasada, prescription 258
zastrzeżenie, clause 70
zawartość, contents 91
zawierać, contain 90
zbierać, collect 74
zbiór, collection 75
zbiór pieśni, songbook 315
zdefektowany, defective 110
zdjęcie, photograph 240
ze, from; on; with 153; 218; 222; 368
zebrać, compile 84
zebranie, assembly 32
zestawiać, compile 84
zeszyt, booklet; number[2] 50; 215
zgromadzenie, assembly 32
zjazd, congress 89
zmieniać, change[1] 63
zmieniać się, change[2] 64
znakowanie, notation 211

źródło, source 316

żurnal, fashion magazine 145
życiorys, biography 48
żywa pagina, running title 292

APPENDIX

styczeń, January	I	**grudzień,** December	XII	**półroczny,** half-yearly,	
luty, February	II	**poniedziałek,** Monday	XIII	semi-annual	XXII
marzec, March	III	**wtorek,** Tuesday	XIV	**kwartalny,** quarterly	XXIII
kwiecień, April	IV	**środa,** Wednesday	XV	**miesiąc,** month	XXIV
maj, May	V	**czwartek,** Thursday	XVI	**miesięczny,** monthly	XXV
czerwiec, June	VI	**piątek,** Friday	XVII	**półmiesięczny,** fort-	
lipiec, July	VII	**sobota,** Saturday	XVIII	nightly, semi-monthly	XXVI
sierpień, August	VIII	**niedziela,** Sunday	XIX	**tydzień,** week	XXVII
wrzesień, September	IX	**rok,** year	XX	**tygodniowy,** weekly	XXVIII
październik, October	X	**roczny,** annual, yearly	XXI	**dzień,** day	XXIX
listopad November	XI			**codzienny,** daily	XXX

NUMERALS — NUMERI

1	**jeden, -dna, -dno**	one	16	**szesnaście**	sixteen
2	**dwa, dwie, dwa**	two	17	**siedemnaście**	seventeen
3	**trzy**	three	18	**osiemnaście**	eighteen
4	**cztery**	four	19	**dziewiętnaście**	nineteen
5	**pięć**	five	20	**dwadzieścia**	twenty
6	**sześć**	six	21	**dwadzieścia jeden**	twenty-one
7	**siedem**	seven	30	**trzydzieści**	thirty
8	**osiem**	eight	40	**czterdzieści**	forty
9	**dziewięć**	nine	50	**pięćdziesiąt**	fifty
10	**dziesięć**	ten	60	**sześćdziesiąt**	sixty
11	**jedenaście**	eleven	70	**siedemdziesiąt**	seventy
12	**dwanaście**	twelve	80	**osiemdziesiąt**	eighty
13	**trzynaście**	thirteen	90	**dziewięćdziesiąt**	ninety
14	**czternaście**	fourteen	100	**sto**	hundred
15	**piętnaście**	fifteen	200	**dwieście**	two hundred

300 trzysta	three hundred		16. szesnasty	sixteenth
400 czterysta	four hundred		17. siedemnasty	seventeenth
500 pięćset	five hundred		18. osiemnasty	eighteenth
600 sześćset	six hundred		19. dziewiętnasty	nineteenth
700 siedemset	seven hundred		20. dwudziesty	twentieth
800 osiemset	eight hundred		21. dwudziesty pierwszy	twenty-first
900 dziewięćset	nine hundred		30. trzydziesty	thirtieth
1000 tysiąc	thousand		40. czterdziesty	fortieth
1. pierwszy, -a, -e	first		60. pięćdziesiąty	fiftieth
2. drugi	second		60. sześćdziesiąty	sixtieth
3. trzeci	third		70. siedemdziesiąty	seventieth
4. czwarty	fourth		80. osiemdziesiąty	eightieth
5. piąty	fifth		90. dziewięćdziesiąty	ninetieth
6. szósty	sixth		100. setny	hundredth
7. siódmi	seventh		200. dwusetny	two hundredth
8. óśmi	eighth		300. trzechsetny	three hundredth
9. dziewiąty	ninth		400. czterechsetny	four hundredth
10. dziesiąty	tenth		500. pięćsetny	five hundredth
11. jedenasty	eleventh		600. sześćsetny	six hundredth
12. dwunasty	twelfth		700. siedemsetny	seven hundredth
13. trzynasty	thirteenth		800. osiemsetny	eight hundredth
14. czternasty	fourteenth		900. dziewięćsetny	nine hundredth
15. piętnasty	fifteenth		1000. tysiączny	thousandth

POLISH ALPHABET
LITTERAE POLONICAE

A a, Ą ą   G g   M m   T t
B b   H h   N n, Ń ń   U u
C c   I i   O o, Ó ó   W w
Ć ć   J j   P p   Y y
D d   K k   R r   Z z
E e, Ę ę   L l   S s   Ż ż
F f   Ł ł   Ś ś   Ż ż

PORTUGUESE—PORTUGALLICA

palavra de título, subject heading	327
panegírico, panegyric	230
papel, paper[1]	231
papel volante, flysheet	148
papiro, papyrus	233
para, for; on; to	150; 222; 344
parágrafo, paragraph	234
parte, part; section	236; 301
partitura, printed music	263
peça, piece	242
pedaço, piece	242
pé de imprensa, imprint	163
pedido, order[2]	225
pequeno, little	188
pergaminho, parchment	235
periódico[1], periodical[1] a.	238
periódico[2], periodical[2] n.	239
plágio, plagiarism	244
poema, poem; poetry	248; 250
poesia, poetry	250
poeta, poet	249
polêmica, polemic	251
poliglota, polyglot	252
por, by; for; from; of; on	55; 150; 153; 218; 222
por baixo de, under	357
pôr em ordem, arrange	29
pós-data, epilogue	135
pós-escrito, postscript	255
póstumo, posthumous	254
prefácio, preface	256
prensa, press[1]	260
preparar, prepare	257
preparar para a impressão, edit[1]	122
prescrição, prescription	258
presidente, president	259
primeira edição, original edition	227
primeiro, first	147
principal, chief; first	66; 147
proémio, preface	256
propriedade intelectual, copyright	98
prosa, prose	267
protocolo, minutes pl.	202
prover, provide (with)	268
provérbio, adage	7
pseudónimo, pseudonym	269
publicação comemorativa, memorial volume	199
publicação finda, ceased publication	61
publicar, edit[2]; publish	123; 271

quotidiano, newspaper	209
raro, rare	275
recolher, summarize	329
recto, recto	277
redacção, editorial office	127
redactor, editor[1]	125
refundir, rewrite	291
refutação, refutation	281
regra, statute[1]	323
reimpressão, reprint[1]	285
repertorio, repertory	282
reprodução, reproduction	286
resenhar, review v.	288
resumir, abridge; summarize	4; 329
resumo, summary	330
retocar, rewrite	291
retrato, portrait	253
reunião, assembly	32
revisão, revision	290
revista, periodical[2] n; review n.	239; 287
romance, novel	213
romance de anticipação científica, science fiction	297
romances e novelas, fiction	146
rosto, title page	343
sair, be published	272
secção, section	301
século, century	62
selecção, selection	303
seleccionar, select	302
sem, without	369
separata, offprint	221
série, series	305
serviço de permuta, exchange centre	138
sessão, sitting	311
s. l. n. d. → sem lugar nem data	
sob, under	357
sobre, sôbre, about; on	3; 222
sociedade, society	313
subscripção, subscription	328
superior, chief	66
suplementário, supplementary	332
suplemento, addenda; appendix; supplement n.	10; 26; 331

APPENDIX

CALENDAR UNITS — TEMPORA

Janeiro, January	I	segunda-feira, Monday	XIII	trimensal, trimestrial,	
Fevereiro, February	II	terça-feira, Tuesday	XIV	quarterly	XXIII
Março, March	III	quarta-feira, Wednesday	XV	mês, month	XXIV
Abril, April	IV	quinta-feira, Thursday	XVI	mensal, monthly	XXV
Maio, May	V	sexta-feira, Friday	XVII	bimensal, quinzenal, fort-	
Junho, June	VI	sábado, Saturday	XVIII	nightly, semi-monthly	XXVI
Julho, July	VII	domingo, Sunday	XIX	semana, week	XXVII
Agosto, August	VIII	ano, year	XX	semanal, weekly	XXVIII
Setembro, September	IX	anual, yearly, annual	XXI	dia, day	XXIX
Outubro, October	X	semestral, half-yearly,		diário, quotidiano, daily	XXX
Novembro, November	XI	semi-annual	XXII		
Dezembro, December	XII				

NUMERALS — NUMERI

1 um, uma	one	16 dezesseis	sixteen
2 dois, duas	two	17 dezessete	seventeen
3 três	three	18 dezoito	eighteen
4 quatro	four	19 dezenove	nineteen
5 cinco	five	20 vinte	twenty
6 seis	six	21 vinte e um	twenty one
7 sete	seven	30 trinta	thirty
8 oito	eight	40 quarenta	forty
9 nove	nine	50 cinquenta	fifty
10 dez	ten	60 sessenta	sixty
11 onze	eleven	70 setenta	seventy
12 doze	twelve	80 oitenta	eighty
13 treze	thirteen	90 noventa	ninety
14 catorze	fourteen	100 cem, cento	hundred
15 quinze	fifteen	200 duzentos	two hundred

300 trezentos	three hundred	16. décimo sexto	sixteenth
400 quatrocentos	four hundred	17. décimo sétimo	seventeenth
500 quinhentos	five hundred	18. décimo oitavo	eighteenth
600 seiscentos	six hundred	19. décimo nono	nineteenth
700 setecentos	seven hundred	20. vigésimo	twentieth
800 oitocentos	eight hundred	21. vigésimo primeiro	twenty-first
900 novecentos	nine hundred	30. trigésimo	thirtieth
1000 mil	thousand	40. quadragésimo	fortieth
1. primeiro	first	50. quinquagésimo	fiftieth
2. segundo	second	60. sexagésimo	sixtieth
3. terceiro	third	70. septuagésimo	seventieth
4. quarto	fourth	80. octagésimo	eightieth
5. quinto	fifth	90. nonagésimo	ninetieth
6. sexto	sixth	100. centésimo	hundredth
7. sétimo	seventh	200. ducentésimo	two hundredth
8. oitavo	eighth	300. trecentésimo	three hundredth
9. nono	ninth	400. quadringentésimo	four hundredth
10. décimo	tenth	500. quingentésimo	five hundredth
11. undécimo, onzeno, décimo primeiro	eleventh	600. sexcentésimo	six hundredth
12. duodécimo, décimo segundo	twelfth	700. septingentésimo	seven hundredth
13. décimo terceiro	thirteenth	800. octingentésimo	eight hundredth
14. décimo quarto	fourteenth	900. nongentésimo, noningentésimo	nine hundredth
15. décimo quinto	fifteenth	1000. milésimo	thousandth

PORTUGUESE ALPHABET

LITTERAE PORTUGALLICAE

A a	I i	Q q
B b	J j	R r
C c	L l	S s
D d	M m	T t
E e, É é	N n	U u
F f	O o	V v
G g	P p	X x
H h		Z z

RUMANIAN–RUMENICA

de, by; from; of	55; 153; 218
decret, order[1]	224
dedicat, dedicated	108
dedicație, dedication	109
defect, defective	110
defectat, defective	110
desen, drawing	120
despre, about; on	3; 222
deziderat, desiderata	113
dialog, dialog	114
dicționar, dictionary	116
din, from; of	153; 218
disc, record[1]	276
diserta, treat	351
disertație, thesis; treatise	340; 352
document, document	117
documentație, documentation	118
dos, back	40
dramă, play	246
drept de autor, copyright	98
dubios, doubtful	119
dublet, duplicate	121
după, after	12
duplicat, copy[1]	96
edita, edit[2]; publish	123; 271
editor, editor[2]; publisher[1]	126; 273
editură, publishing house	274
ediție, edition	124
ediție completă, complete works	87
ediție festivă, memorial volume	199
ediție omagială, memorial volume	199
ediție princeps, original edition	227
elabora, elaborate	128
elegie, elegy	129
elogiu, panegyric	230
enciclopedie, encyclopaedia	130
epilog, epilogue	135
epistolă, letter	181
epitalam, bridal song	53
epopee, epic	134
eseu, essay[1]	136
exemplar, copy[2]	97
exemplar de obligație, deposit copy	112
explica, explain	140
explicație, explanation	141
expoziție, exhibition	139
extras, extract; offprint	142; 221

facsimil, facsimile	144
falsificare, forgery	151
falsificație, forgery	151
fasciculă, number[2]	215
față, recto	277
fără, without	369
figură, picture	241
filă, leaf	178
fișă, card	56
fișă de catalog, card	56
fișă de trimitere, reference card	278
fișă principală, main card	191
fișă provizorie, temporary card	336
fișă suplimentară/secundară, added entry	9
f. l. ș. a. → fără loc și an	
foaie, leaf; sheet	178; 307
foaie volantă, flysheet	148
fotografie, photograph	240
fototipie, collotype	76
fracțiune, fragment	152
fragment, fragment	152
frontispiciu, title page	343
gazetă, newspaper	209
general, chief	66
ghid, guide(-book)	155
glosar, glossary	154
grava, engrave	131
gravură, copperplate engraving, engraving	95; 132
gravură cu acvaforte, etching	137
gravură în lemn, woodcut	370
gravură pe cupru, copperplate	94
greșeală de tipar, printer's error	265
hartă, map	193
hîrtie, paper[1]	231
-i pl., the	339
iconografie, iconography	158
idilă, idyll	159
ieși de sub tipar, be published	272
ilustra, illustrate; represent	160; 284
ilustrație, picture	241
imagine, picture	241
imitație, imitation	161
imn, anthem	24
imprima, print[1]	261
imprimare, impression	162

APPENDIX

CALENDAR UNITS — TEMPORA

ianuarie, January	I	**luni,** Monday	XIII	**trimestrial,** quarterly	XXIII
februarie, February	II	**marți,** Tuesday	XIV	**lună,** month	XXIV
martie, March	III	**miercuri,** Wednesday	XV	**lunar,** monthly	XXV
aprilie, April	IV	**joi,** Thursday	XVI	**bilunar,** fortnightly,	
mai, May	V	**vineri,** Friday	XVII	semi-monthly	XXVI
iunie, June	VI	**sîmbătă,** Saturday	XVIII	**săptămînă,** week	XXVII
iulie, July	VII	**duminică,** Sunday	XIX	**săptămînal, ebdomadar,**	
august, August	VIII	**an,** year	XX	weekly	XXVIII
septembrie, September	IX	**anual,** annual, yearly	XXI	**zi,** day	XXIX
octombrie, October	X	**bianual, semestrial,** half-		**zilnic, cotidian,** daily	XXX
noiembrie, November	XI	yearly, semi-annual	XXII		
decembrie, December	XII				

NUMERALS — NUMERI

1 **un, unu; o, una**	one	16 **șaisprezece**	sixteen
2 **doi, două**	two	17 **șaptesprezece**	seventeen
3 **trei**	three	18 **optsprezece**	eighteen
4 **patru**	four	19 **nouăsprezece**	nineteen
5 **cinci**	five	20 **douăzeci**	twenty
6 **șase**	six	21 **douăzeci și unu/una**	twenty one
7 **șapte**	seven	30 **treizeci**	thirty
8 **opt**	eight	40 **patruzeci**	forty
9 **nouă**	nine	50 **cin(ci)zeci**	fifty
10 **zece**	ten	60 **șaizeci**	sixty
11 **unsprezece**	eleven	70 **șaptezeci**	seventy
12 **doisprezece, douăsprezece**	twelve	80 **optzeci**	eighty
13 **treisprezece**	thirteen	90 **nouăzeci**	ninety
14 **paisprezece, patrusprezece**	fourteen	100 **o sută**	hundred
15 **cin(ci)sprezece**	fifteen	200 **două sute**	two hundred

300 trei sute	three hundred	
400 patru sute	four hundred	
500 cinci sute	five hundred	
600 şase sute	six hundred	
700 şapte sute	seven hundred	
800 opt sute	eight hundred	
900 nouă sute	nine hundred	
1000 o mie	thousand	

1. întîi(ul), primul; întîi(a), prima	first	16. al şaisprezecelea	sixteenth
2. al doilea, a doua	second	17. al şaptesprezecelea	seventeeth
3. al treilea, a treia	third	18. al optsprezecelea	eighteenth
4. al patrulea, a patra	fourth	19. al nouăsprezecelea	nineteenth
5. al cincilea, a cincea	fifth	20. al douăzecilea, a douăzecea	twentieth
6. al şaselea, a şasea	sixth	21. al douăzeci şi unulea	twenty first
7. al şaptelea, a şaptea	seventh	30. al treizecilea	thirtieth
8. al optulea, a opta	eighth	40. al patruzecilea	fortieth
9. al nouălea, a noua	ninth	50. al cincizecilea	fiftieth
10. al zecelea, a zecea	tenth	60. al şaizecilea	sixtieth
11. al unsprezecelea, a unsprezecea	eleventh	70. al şaptezecilea	seventieth
12. al doisprezecelea	twelfth	80. al optzecilea	eightieth
13. al treiprezecelea	thirteenth	90. al nouăzecilea	ninetieth
14. al paisprezecelea/patrusprezecelea	fourteenth	100. al sutălea, a suta	hundredth
15. al cin(ci)sprezecelea	fifteenth	200. al două sutălea, a două suta	two hundredth
		300. al trei sutălea	three hundredth
		400. al patru sutălea	four hundredth
		500. al cinci sutălea	five hundredth
		600. al şase sutălea	six hundredth
		700. al şapte sutălea	seven hundredth
		800. al opt sutălea	eight hundredth
		900. al nouă sutălea	nine hundredth
		1000. al miilea, a mia	thousandth

RUMANIAN ALPHABET

LITTERAE RUMENICAE

A a â	Î î	S s
Ă ă	J j	Ş ş
B b	K k	T t
C c	L l	Ţ ţ
D d	M m	U u
E e	N n	V v
F f	O o	W w
G g	P p	X x
H h	Q q	Z z
I i	R r	

SERBIAN – SERVICA

izdavanje (издавање), edition 124
izdavati (издавати), publish 271
izlaziti (излазити), *be* published 272
izložba (изложба), exhibition 139
izreka (изрека), adage 7
izvadak (извадак), extract 142
izveštaj (извештај), report[1]; transactions *pl.* 346
izvod (извод), extract 142
izvor (извор), source 316

javna biblioteka (јавна библиотека), public library 270
jedan, jedna, jedno (један, једна, једно), a, an 1
jedini (једини), only *a.* 223
jezik (језик), language 175

k (к), ka (ка) to, 344
kamenotisak (каменотисак), lithograph(y) 187
katalog (каталог), catalog(ue) *n.* 58
katalogizacija (каталогизацја), cataloging 60
katalogizovati (каталогизовати), catalog(ue) *v.* 59
katalog na listićima (каталог на листићима), card index/catalog(ue) 57
kataloški listić (каталошки листић), card 56
kazalo (казало), table of contents 334
kitica (китица), strophe 325
klasifikacija (класификација), classification 69
klauzula (клаузула), clause 70
knjiga (књига), book 49
knjiga sa slobodnim listovima (књига са слободним листовима), loose-leaf book 190
knjiga za decu (књига за децу), children's book 67
knjigoveznica (књиговезница), bindery 46
knjižara (књижара), bookshop 51
književnik (књижевник), writer 374
književnost (књижевност), literature 186
knjižnica (књижница), library 183
kodeks (кодекс), codex 71
kolektivni autor (колекⰰивни аутор), corporate author 100
kolona (колона), column 77
komad (комад), piece 242
komedija (комедија), comedy 78
komentar (коментар), commentary 82
komentarisati (коментарисати), comment[1] *v.* 81
komisija (комисија), committee 83
konferencija (конференција), conference 88

kongres (конгрес), congress 89
kopija (копија), copy[1] 96
korektura (коректура), correction 102
korespondencija (кореспонденција), correspondence 103
koričiti (коричити), bind 45
koristiti (користити), use *v.* 363
korišćenje (коришћење), use *n.* 362
kritika (критика), critique 104
kritikovati (критиковати), criticize 105
krivotvorina (кривотворина), forgery 151
kserografija (ксерографија), xerography 375

lekcija (лекција), lecture 179
letak (летак), flysheet 148
leto (лето), year 376
letopisi (летописи) *pl.*, annals *pl.* 19
list (лист), leaf; letter; newspaper 178; 181; 209
literatura (литература), literature 186
litografija (литографија), lithograph(y) 187

magazin za knjige (магазин за књиге), stack-room 320
na magnetofonskoj traci (на магнетофонској траци), on tape 335
mali (мали), little 188
margina (маргина), margin 194
menjati (мењати), change[1] 63
menjati se (мењати се), change[2] 64
memoari (мемоари) *pl.*, memoirs *pl.* 197
memorandum (меморандум), memoir 196
bez mesta i godine, b. m. i g. (без места и године, б. м. и г.), no place no date 210
mesto (место), place 243
mešovit (мешовит), miscellanea *pl.* 203
mikrofilm (микрофилм), microfilm 200
minijatura (минијатура), miniature 201
mnogojezičan (многојезичан), polyglot 252
modni žurnal (модни журнал), fashion magazine 145
monografija (монографија), monograph(y) 204
mudra izreka (мудра изрека), adage 7
muzika (музика), music 205
muzikalije (музикалије) *pl.*, printed music 263

na (на), in; on 164; 222
naličje (наличје), verso 365
napomena (напомена), annotation; note[1] 22; 212
naredba (наредба), order[1] 224

narodna biblioteka (народна библиотека),
  national library   207
narudžbina (наруџбина), order [2]   225
nasledstvo (наследство), literary remains *pl.*   185
naslov (наслов), title   341
naslovna strana (насловна страна), title page   343
nastavak (наставак), sequel   304
nastavljati (настављати), continue   92
natpis (натпис), title   341
naučno-fantastični roman (научно-фантастични ро-
  ман), science fiction   297
nauka (наука), science   296
neizdan (неиздан), unpublished   361
nekompletan (некомплетан), incomplete   165
nekrolog (некролог), obituary notice   216
nepotpun (непотпун), defective   110
nepromenljiv (непроменљив), unchanged   356
neverodostojan (неверодостојан), doubtful   119
normiranje (нормирање), standardisation   321
nov (нов), new   208
novela (новела), short story   309
novine (новине) *pl.*, newspaper   209

o (о), about; on   3; 222
obavezan primerak (обавезан примерак), deposit
  copy   112
objašnjavati (објашњавати), explain   140
objašnjenje (објашњење), commentary; explanation
     82; 141
objavljivati (објављивати), *be* published   272
obraditi (обрадити), elaborate   128
ocenjivati (оцењивати), criticize   105
od (од), by; from; of   55; 153; 218
odabrati (одабрати), select   302
odbor (одбор), committee   301
odeljak (одељак), section   301
odlomak (одломак), extract; fragment   142; 152
odobriti (одобрити), authorize   37
ogled (оглед), essay[1]; study   136; 326
opaska (опаска), annotation; note[1]   22; 212
opovrgavanje (оповргавање), refutation   281
originalan (оригиналан), original   226
originalni rukopis (оригинални рукопис), autograph   39
na osnovu ... (на основу ...), based on the...   41
otisak (отисак), impression; printed matter   162; 262
ovde-onde (овде-онде), passim   237
ovlastiti (овластити), authorize   37

pa (па), and   18
paginacija (пагинација), paging   229
papir (папир), paper[1]   231
papirus (папирус), papyrus   233
paragraf (параграф), paragraph   234
parče (парче), piece   242
pergament (пергамент), parchment   235
periodičan (периодичан), periodical[1] *a.*   238
personalna bibliografija (персонална библиографија), author bibliography   36
pesma (песма), poem; song   248; 314
pesmarica (песмарица), songbook   315
pesnik (песник), poet   249
pesništvo (песништво), poetry   250
pisac (писац), writer   374
pisanje (писање), script   299
pisati (писати), write   372
pisati pesme (писати песме), write poetry   373
pismo (писмо), letter   181
plagiat (плагиат), plagiarism   244
pod (под), about; on; under   3; 222; 357
podaci (подаци) *pl.*, contribution   93
podražavanje (подражавање), imitation   161
poezija (поезија), poetry   250
poglavlje (поглавље), chapter   65
pogovor (поговор), epilogue   135
pohvalna pesma (похвална песма), panegyric   230
polemičan spis (полемичан спис), polemic   251
popis (попис), list   184
popraviti (поправити), correct   101
popularna biblioteka (популарна библиотека), public library   270
popuniti (попунити), complete *v.*   86
portret (портрет), portrait   253
poseban otisak (посебан отисак), offprint   221
posle (после), after   12
posmrtni (посмртни), posthumous   254
posmrtni govor (посмртни говор), memorial speech   198
postskriptum (постскриптум), postscript   255
posvećen (посвећен), dedicated   108
posveta (посвета), dedication   109
potpun (потпун), complete *a.*   85
povest (повест), history   157
povez (повез), binding   47
povezati (повезати), bind   45
pozajmica (позајмица), lending   180
pozorišni komad (позоришни комад), play   246
pravi (прави), authentic   34

serija (серија), series   305
signatura (сигнатура), location mark; signature 189; 310
skica (скица), sketch   312
skraćeni opis (скраћени опис), abbreviated entry   2
skraćivati (скраћивати), abridge   4
skupljati (скупљати), collect   74
slagati po abecedi (слагати по абецеди), alphabetize   16
slika (слика), picture   241
slog (слог), matter   195
složiti (сложити), compile   84
služba razmene publikacija (служба размене публикација), exchange centre   138
služben (службен), official   220
snabdeti (снабдети), provide (with)   268
specijalna biblioteka (специјална библиотека), special library   318
spisak (списак), list   184
spomenica (споменица), memoir; memorial volume   196; 199
spomen-knjiga (спомен-књига), memorial volume   199
spomen-slovo (спомен слово), memorial speech   198
sporedna kataloška jedinica (споредна каталошка јединица), added entry   9
standardizacija (стандардизација), standardisation   321
star (стар), ancient   17
statut (статут), statute   323
stenografija (стенографија), shorthand   308
stih (стих), poem   248
stoleće (столеће), century   62
stranica (страница), page   228
strip (стрип), comic strip   80
strofa (строфа), strophe   325
stručna bibliografija (стручна библиографија), special bibliography   317
studija (студија), study   326
stvarni naslov (стварни наслов), title entry   342
suautor (суаутор), joint author   174
sumnjiv (сумњив), doubtful   119
svadbena pesma (свадбена песма), bridal/nuptial song   53
sveska (свеска), booklet; number[2]; volume[1] 50; 215; 366
svezak (свезак), volume[1]   366

škola (школа), school   295
štampa (штампа), impression; press[1]   162; 260
štampar (штампар), printer   264
štamparija (штампарија), printing office   266

štamparska greška (штампарска грешка), printer's error   265
štamparstvo (штампарство), typography   355
štampati (штампати), print[1]   261
štititi (штитити), defend   111

tabak (табак), sheet   307
tabela (табела), plate; table   245; 333
tablica (таблица), plate   245
tekst (текст), text   337
tekući naslov (текући наслов), running title   292
tipografija (типографија), printing office   266
tipovi slova (типови слова) pl., type size   354
tisak (тисак), press[1]   260
tiskanica (тисканица), printed matter   262
tiskara (тискара), typography   355
tiskati (тискати), print[1]   261
titula (титула), title   341
tom (том), volume[1]   366
tragedija (трагедија), tragedy   345
traktat (трактат), treatise   352
transkripcija (транскрипција), transcription   347
transliteracija (транслитерација), transliteration   350
tumačiti (тумачити), explain; translate   140; 348
tvorevina (творевина), work   371

u (у), in   164
udžbenik (уџбеник), textbook   338
ukratko izložiti (укратко изложити), summarize   329
umetati (уметати), insert   169
umetnost (уметност), art   30
umnožiti (умножити), enlarge   133
univerzitet (универзитет), university   359
univerzitetska biblioteka (универзитетска библиотека), university library   360
upotreba (употреба), use n.   362
upotrebljavati (употребљавати), use v.   363
uputni kataloški listić (упутни каталошки листић), reference card   278
ured (уред), office   219
uredba (уредба), order[1]   224
urezivati (урезивати), engrave   131
uvod (увод), introduction   172
uvoditi (уводити), introduce   171

varijant (варијант), version   364
vek (век), century   62

APPENDIX

CALENDAR UNITS — TEMPORA

januar (jануар), January	I
februar (фебруар), February	II
mart (март), March	III
april (април), April	IV
maj (мај), May	V
juni (јуни), June	VI
juli (јули), July	VII
avgust (август), August	VIII
septembar (септембар), September	IX
oktobar (октобар), October	X
novembar (новембар), November	XI
decembar (децембар), December	XII
ponedeljak (понедељак), Monday	XIII
utorak (уторак), Tuesday	XIV
sreda (среда), Wednesday	XV
četvrtak (четвртак), Thursday	XVI
petak (петак), Friday	XVII
subota (субота), Saturday	XVIII
nedelja (недеља), Sunday	XIX
godina (година), year	XX
godišnji (годишњи), annual, yearly	XXI
polugodišnji (полугодишњи), half-yearly, semiannual	XXII
tromesečni (тромесечни), quarterly	XXIII
mesec (месец), month	XXIV
mesečni (месечни), monthly	XXV
polumesečni (полумесечни), fortnightly, semi-monthly	XXVI
nedelja, sedmica (недеља, седмица), week	XXVII
sedmični (седмични), weekly	XXVIII
dan (дан), day	XXIX
dnevni (дневни), daily	XXX

NUMERALS — NUMERI

1 **jedan, -dna, -dno** (један, -дна, -дно)	one
2 **dva, dve** (два, две)	two
3 **tri** (три)	three
4 **četiri** (четири)	four
5 **pet** (пет)	five
6 **šest** (шест)	six
7 **sedam** (седам)	seven
8 **osam** (осам)	eight
9 **devet** (девет)	nine
10 **deset** (десет)	ten
11 **jedanaest** (једанаест)	eleven
12 **dvanaest** (дванаест)	twelve
13 **trinaest** (тринаест)	thirteen

14	četrnaest (четрнаест)	fourteen	6.	šesti (шести)	sixth
15	petnaest (петнаест)	fifteen	7.	sedmi (седми)	seventh
16	šesnaest (шеснаест)	sixteen	8.	osmi (осми)	eighth
17	sedamnaest (седамнаест)	seventeen	9.	deveti (девети)	ninth
18	osamnaest (осамнаест)	eighteen	10.	deseti (десети)	tenth
19	devetnaest (деветнаест)	nineteen	11.	jedanaesti (једанаести)	eleventh
20	dvadeset (двадесет)	twenty	12.	dvanaesti (дванаести)	twelfth
21	dvadeset i jedan (двадесет и један)	twenty-one	13.	trinaesti (тринаести)	thirteenth
30	trideset (тридесет)	thirty	14.	četrnaesti (четрнаести)	fourteenth
40	četrdeset (четрдесет)	forty	15.	petnaesti (петнаести)	fifteenth
50	pedeset (педесет)	fifty	16.	šesnaesti (шеснаести)	sixteenth
60	šezdeset (шездесет)	sixty	17.	sedamnaesti (седамнаести)	seventeenth
70	sedamdeset (седамдесет)	seventy	18.	osamnaesti (осамнаести)	eighteenth
80	osamdeset (осамдесет)	eighty	19.	devetnaesti (деветнаести)	nineteenth
90	devedeset (деведесет)	ninety	20.	dvadeseti (двадесети)	twentieth
100	sto, stotina (сто, стотина)	hundred	21.	dvadeset i prvi (двадесет и први)	twenty-first
200	dve stotine, dvesta (две стотине, двеста)	two hundred	30.	trideseti (тридесети)	thirtieth
300	tri stotine, trista (три стотине, триста)	three hundred	40.	četrdeseti (четрдесети)	fortieth
400	četiri stotine, četiristo (четири стотине, четиристо)	four hundred	50.	pedeseti (педесети)	fiftieth
500	petsto, pet stotina (петсто, пет стотина)	five hundred	60.	šezdeseti (шездесети)	sixtieth
600	šeststo (шестсто)	six hundred	70.	sedamdeseti (седамдесети)	seventieth
700	sedamsto (седамсто)	seven hundred	80.	osamdeseti (осамдесети)	eightieth
800	osamsto (осамсто)	eight hundred	90.	devedeseti (деведесети)	ninetieth
1900	devetsto (деветсто)	nine hundred	100.	stoti (стоти)	hundredth
000	hiljada (хиљада)	thousand	200.	dvstoti (двстоти)	two hundredth
1.	prvi, -a, -o (први, -а, -о)	first	300.	tristoti (тристоти)	three hundredth
2.	drugi (други)	second	400.	četiristoti (четиристоти)	four hundredth
3.	treći (трећи)	third	500.	petstoti (петстоти)	five hundredth
4.	četvrti (четврти)	fourth	600.	šeststoti (шестстоти)	six hundredth
5.	peti (пети)	fifth	700.	sedamstoti (седамстоти)	seven hundredth
			800.	osamstoti (осамстоти)	eight hundredth
			900.	devetstoti (деветстоти)	nine hundredth
			1000.	hiljaditi, tisući (хиљадити, тисући)	thousandth

SERBIAN (CYRILLIC) ALPHABET. TRANSLITERATION
LITTERAE SERVICAE (CYRILLIANAE). TRANSLITTERATIO

А а,	a	Ј ј,	j	Т т,	t
Б б,	b	К к,	k	Ћ ћ,	ć
В в,	v	Л л,	l	У у,	u
Г г,	g	М м,	m	Ф ф,	f
Д д,	d	Н н,	n	Х х,	h
Ђ ђ,	đ (dj)	Њ њ,	nj	Ц ц,	c
Е е,	e	О о,	o	Ч ч,	č
Ж ж,	ž	П п,	p	Џ џ,	dž
З з,	z	Р р,	r	Ш ш,	š
И и,	i	С с,	s		

SLOVAK – SLOVACA

APPENDIX

CALENDAR UNITS — TEMPORA

január, January	I	**december,** December	XII	**polročný,** half-yearly,	
február, February	II	**pondelok,** Monday	XIII	semi-annual	XXII
marec, March	III	**utorok,** Tuesday	XIV	**štvťročný,** quarterly	XXIII
apríl, April	IV	**streda,** Wednesday	XV	**mesiac,** month	XXIV
máj, May	V	**štvrtok,** Thursday	XVI	**mesačný,** monthly	XXV
jún, June	VI	**piatok,** Friday	XVII	**polmesačný,** fortnightly,	
júl, July	VII	**sobota,** Saturday	XVIII	semi-monthly	XXVI
august, August	VIII	**nedeľa,** Sunday	XIX	**týždeň,** week	XXVII
september, September	IX	**rok,** year	XX	**týždňový,** weekly	XXVIII
október, October	X	**ročný,** yearly, annual	XXI	**deň,** day	XXIX
november, November	XI			**denný, dňový,** daily	XXX

NUMERALS — NUMERI

1	**jeden, -dna, -dno**	one	16	**šestnásť**	sixteen
2	**dva, dve**	two	17	**sedemnásť**	seventeen
3	**tri**	three	18	**osemnásť**	eighteen
4	**štyri**	four	19	**devätnásť**	nineteen
5	**päť**	five	20	**dvadsať**	twenty
6	**šesť**	six	21	**dvadsaťjeden,**	
7	**sedem**	seven		**jedenadvadsať**	twenty one
8	**osem**	eight	30	**tridsať**	thirty
9	**deväť**	nine	40	**štyridsať**	forty
10	**desať**	ten	50	**päťdesiat**	fifty
11	**jedenásť**	eleven	60	**šesťdesiat**	sixty
12	**dvanásť**	twelve	70	**sedemdesiat**	seventy
13	**trinásť**	thirteen	80	**osemdesiat**	eighty
14	**štrnásť**	fourteen	90	**deväťdesiat**	ninety
15	**pätnásť**	fifteen	100	**sto**	hundred

200	dvesto	two hundred	
300	tristo	three hundred	
400	štyristo	four hundred	
500	päťsto	five hundred	
600	šesťsto	six hundred	
700	sedemsto	seven hundred	
800	osemsto	eight hundred	
900	deväťsto	nine hundred	
1000	tisíc	thousand	
1.	prvý, -á, -é	first	
2.	druhý	second	
3.	tretí	third	
4.	štvrtý	fourth	
5.	piaty	fifth	
6.	šiesty	sixth	
7.	siedmy	seventh	
8.	ôsmy	eighth	
9.	deviaty	ninth	
10.	desiaty	tenth	
11.	jedenásty	eleventh	
12.	dvanásty	twelfth	
13.	trinásty	thirteenth	
14.	štrnásty	fourteenth	
15.	pätnásty	fifteenth	

16.	šestnásty	sixteenth
17.	sedemnásty	seventeenth
18.	osemnásty	eighteenth
19.	devätnásty	nineteenth
20.	dvadsiaty	twentieth
21.	dvadsiaty prvý, jedenadvadsiaty	twenty first
30.	tridsiaty	thirtieth
40.	štyridsiaty	fortieth
50.	päťdesiaty	fiftieth
60.	šesťdesiaty	sixtieth
70.	sedemdesiaty	seventieth
80.	osemdesiaty	eightieth
90.	deväťdesiaty	ninetieth
100.	stý	hundredth
200.	dvojstý	two hundredth
300.	trojstý	three hundredth
400.	štvorstý	four hundredth
500.	päťstý	five hundredth
600.	šesťstý	six hundredth
700.	sedemstý	seven hundredth
800.	osemstý	eight hundredth
900.	deväťstý	nine hundredth
1000.	tisíci	thousandth

SLOVAK ALPHABET
LITTERAE SLOVACAE

A a, Á á, ä B b C c Č č D d, Ď ď E e, É é F f G g H h

Ch ch I i, Í í J j K k L l, Ľ ľ M m N n, Ň ň O o, Ó ó, Ô ô P p Q q

R r, ŕ S s Š š T t, Ť ť U u, Ú ú V v W w X x Y y, ý Z z Ž ž

375

SWEDISH — SUECICA

APPENDIX

CALENDAR UNITS — TEMPORA

januari, January	I	**december,** December	XII	**halvårs-,** half-yearly,		
februari, February	II	**måndag,** Monday	XIII	semi-annual	XXII
mars, March	III	**tisdag,** Tuesday	XIV	**kvartals-,** quarterly	XXIII
april, April	IV	**onsdag,** Wednesday	XV	**månad,** month	XXIV
maj, May	V	**torsdag,** Thursday	XVI	**månatlig,** monthly	XXV
juni, June	VI	**fredag,** Friday	XVII	**halvmånads-,** fortnight-	
juli, July	VII	**lördag,** Saturday	XVIII	ly, semi-monthly	XXVI
augusti, August	VIII	**söndag,** Sunday	XIX	**vecka,** week	XXVII
september, September	IX	**år,** year	XX	**vecko-,** weekly	XXVIII
oktober, October	X	**årlig,** annual, yearly	XXI	**dag,** day	XXIX
november, November	XI			**daglig,** daily	XXX

NUMERALS — NUMERI

1 **en, ett**	one	16 **sexton**	sixteen
2 **två**	two	17 **sjutton**	seventeen
3 **tre**	three	18 **aderton, arton**	eighteen
4 **fyra**	four	19 **nitton**	nineteen
5 **fem**	five	20 **tjugo, tjugu**	twenty
6 **sex**	six	21 **tjugoen, tjugoett**	twenty-one
7 **sju**	seven	30 **trettio**	thirty
8 **åtta**	eight	40 **fyrtio**	forty
9 **nio**	nine	50 **femtio**	fifty
10 **tio**	ten	60 **sextio**	sixty
11 **elva**	eleven	70 **sjuttio**	seventy
12 **tolv**	twelve	80 **åttio**	eighty
13 **tretton**	thirteen	90 **nittio**	ninety
14 **fjorton**	fourteen	100 **hundra**	hundred
15 **femton**	fifteen	200 **tvåhundra**	two hundred

300 trehundra	three hundred	
400 fyrahundra	four hundred	
500 femhundra	five hundred	
600 sexhundra	six hundred	
700 sjuhundra	seven hundred	
800 åttahundra	eight hundred	
900 niohundra	nine hundred	
1000 tusen	thousand	
1. första, -e	first	
2. andra, -e	second	
3. tredje	third	
4. fjärde	fourth	
5. femte	fifth	
6. sjätte	sixth	
7. sjunde	seventh	
8. åttonde	eighth	
9. nionde	ninth	
10. tionde	tenth	
11. elfte	eleventh	
12. tolfte	twelfth	
13. trettonde	thirteenth	
14. fjortonde	fourteenth	
15. femtonde	fifteenth	

16. sextonde	sixteenth
17. sjuttonde	seventeenth
18. adertonde, artonde	eighteenth
19. nittonde	nineteenth
20. tjugonde	twentieth
21. tjugoförsta	twenty-first
30. trettionde	thirtieth
40. fyrtionde	fortieth
50. femtionde	fiftieth
60. sextionde	sixtieth
70. sjuttionde	seventieth
80. åttionde	eightieth
90. nittionde	ninetieth
100. hundrade	hundredth
200. tvåhundrade	two hundredth
300. trehundrade	three hundredth
400. fyrahundrade	four hundredth
500. femhundrade	five hundredth
600. sexhundrade	six hundredth
700. sjuhundrade	seven hundredth
800. åttahundrade	eight hundredth
900. niohundrade	nine hundredth
1000. tusende	thousandth

SWEDISH ALPHABET

LITTERAE SUECICAE

A a	K k	U u
B b	L l	V v
C c	M m	W w
D d	N n	X x
E e	O o	Y y
F f	P p	Z z
G g	Q q	Å å
H h	R r	Ä ä
I i	S s	Ö ö
J j	T t	

384

CONTENTS – INDEX